A GUIDE

SITE GU

FOR ELECTRICAL INSTALLATIONS UP TO 100 A

updated to

18TH EDITION

BS 7671:2018

ERTSURE

ertsure LLP is a limited liability partnership between the Electrical Safety First and the Electrical Contractors' ssociation (ECA) registered in England and Wales. Its brands include NICEIC the UK's leading voluntary egulatory body for the electrical contracting industry, which has been assessing the electrical competence of lectricians for over fifty years and currently maintains a roll of over 36,000 registered contractors. It also includes LECSA, who provide inspection, assessment and certification services to contractors working with electrical and enewable technology installations in private dwellings.

ertsure products are delivered through the NICEIC and ELECSA brands. The products include, amongst others:

The NICEIC Approved Contractor Scheme, a scheme for electrical contractors undertaking design, installation, commissioning and maintenance of electrical installations to *BS 7671*.

The ELECSA and NICEIC Part P Domestic Installer Scheme, which registers installers to Full Scope or Defined Competence for domestic electrical installation work in accordance with Part P of the Building Regulations in England and Wales.

The Microgeneration Certification Scheme (MCS), developed to assess businesses installing microgeneration technology and is designed to protect the consumer.

Schemes, for energy saving improvements to dwellings and other buildings.

LECTRICAL SAFETY FIRST

ectrical Safety First (formerly the Electrical Safety Council) is an independent charity committed to reducing eaths and injuries through electrical accidents at home and at work.

ectrical Safety First is supported by all sectors of the electrical industry, approvals and research bodies, onsumer interest organisations, the electricity distribution industry, professional institutes and institutions, egulatory bodies, trade and industry associations and federations, trade unions and local and central overnment.

HE ELECTRICAL CONTRACTORS' ASSOCIATION

ne Electrical Contractors' Association (ECA) is the leading trade association for contractors involved in electrical ngineering projects in England, Northern Ireland and Wales.

ublished by:

ertsure LLP, Warwick House, Houghton Hall Park, Houghton Regis, Dunstable, Bedfordshire LU5 5ZX

el: 0333 015 6626
ebsites: www.certsure.com www.niceic.com www.elecsa.co.uk www.shop.niceic.com

BN: 978-1-906091-91-0 EAN: 9781906091910 Sap code: PNICSG18

blished July 2018

CONTENTS

CHAPTER 1
INTRODUCTION

1.1 Scope

This Guide is intended for contractors who carry out electrical installation work in dwellings, including houses and flats. It provides information and guidance relating to the design, construction, inspection, testing and certification of domestic electrical installation work, and is intended to promote good practice and aid understanding of the fundamental concepts of protection for safety. The guidance is based on the requirements of *BS 7671: 2018 Requirements for Electrical Installations*, and references to Regulation numbers are to those contained in that Standard.

1.2 Statutory and non-statutory requirements

1.2.1 Health and Safety at Work etc Act 1974

The Health and Safety at Work etc Act 1974 (HSW Act) is the primary legislation dealing with occupational health and safety in the United Kingdom. This is an enabling Act, a provision in law that confers on appropriate officials – in this case the Health and Safety Executive – the power to implement or enforce the law. The HSW Act sets out the general duties that employers have towards employees and members of the public, and that employees have to themselves and to each other.

The principle of *'so far as is reasonably practicable'* may be applied to some of these duties. This principle is considered further in Section 1.3.1 of this Guide along with the principle of absolute duty. In the case of offences under the HSW Act, a person is deemed 'guilty' until proven 'innocent' which is the opposite of the presumption of innocence that exists in 'criminal' law. This means that a person charged with a health and safety offence must demonstrate that they took the relevant precautions in relation to the duties placed on them by the HSW Act.

1.2.2 Electricity at Work Regulations 1989

The Electricity at Work Regulations 1989 (EWR) is a Statutory Instrument made under the Health and Safety at Work etc. Act 1974 (HSW Act).

A statutory instrument is a form of subordinate or secondary legislation, which is made under an enabling Act.

Statutory instruments are classified according to their subject matter as being local or general in application. The requirements contained in the EWR are generally applicable. That is to say, they apply (with very few exceptions) wherever work is carried out on, or close to, live conductors.

The provisions contained within a statutory instrument must be complied with; they are not optional or preferred conditions.

EWR were introduced to require precautionary measures to be taken against the risk of death or personal injury caused by electricity when carrying out work activities.

EWR contains 21 regulations. A brief summary of the general regulations is provided below.

Regulation 2 defines a number of terms used within the Regulations, including *danger; electrical equipment; and injury.*

Regulation 3 requires employers and employees to comply with 'the provisions contained within EWR in so far as they relate to matters within (their) control'. The nature of these duties is expanded on in the Health and Safety Executive (HSE) publication *The Electricity at Work Regulations 1989 - Guidance on Regulations* (HSR25), which is considered further in Section 1.3.1 of this chapter.

Regulation 4 gives requirements relating to the construction, use and maintenance of systems, and the electrical equipment therein, and is reproduced below:

1) All systems shall at all times be of such construction as to prevent, so far as is reasonably practicable, danger.
2) As may be necessary to prevent danger, all systems shall be maintained so as to prevent, so far as is reasonably practicable, such danger.
3) Every work activity, including operation, use and maintenance of a system and work near a system, shall be carried out in such a manner as not to give rise, so far as is reasonably practicable, to danger.
4) Any equipment provided under these Regulations for the purpose of protecting persons at work on or near electrical equipment shall be suitable for the use for which it is provided, be maintained in a condition suitable for that use, and be properly used.

Regulation 5 is concerned with the strength and capability of equipment with regard to its use under normal, transient and prospective fault conditions.

Regulation 6 relates to the construction and protection of electrical equipment against the adverse or hazardous environments to which electrical equipment might reasonably be exposed in use. This is of relevance to the use of the International Protection (IP) code described in *BS EN 60529: 1992* (as amended).

Regulation 7 requires conductors that may give rise to danger to be covered by insulation and where necessary be provided with other means of protection; or be so placed to prevent so far as is reasonably practicable, danger.

Regulation 8 is concerned with the provision of earthing or other precautions to prevent danger.

Regulation 9 deals with the integrity (that is, the continuity) of referenced conductors.

Regulation 10 requires all joints and connections to be mechanically and electrically suitable for use.

Regulation 11 requires means of protecting against excess of current to be provided.

Regulation 12 calls for the provision of secure means of isolation to any electrical equipment – that is, whole installations, or one or more circuits, or individual items as necessary to prevent danger.

Regulation 13 states that:

> Adequate precautions shall be taken to prevent electrical equipment, which has been made dead in order to prevent danger while work is carried out on or near that equipment, from becoming electrically charged during that work if danger may thereby arise.

Regulation 14 prescribes three conditions which must all be satisfied before a person may be involved in work on or near live conductors. The conditions are:

a) it is unreasonable in all the circumstances for it to be dead; and

b) it is reasonable in all the circumstances for him to be at work on or near it while it is live; and

c) suitable precautions (including where necessary the provision of suitable protective equipment) are taken to prevent injury.

Regulation 15 requires the provision of adequate working space, means of access, and lighting where electrical equipment is situated.

Regulation 16 requires persons to be competent to prevent danger or injury and is worthy of repeat here:

> No person shall be engaged in any work activity where technical knowledge or experience is necessary to prevent danger or, where appropriate, injury unless he possesses such knowledge or experience, or is under such degree of supervision as may be appropriate having regard to the nature of the work.

The issue of competency is considered further in Section 1.3.1 of this chapter.

1.2.3 Construction (Design and Management) Regulations 2015

The current edition of the Construction (Design and Management) (CDM 2015) Regulations came into force in April 2015. The Regulations place a legal duty on everyone involved in construction work from the design stage through to completion to:

- sensibly plan the work so the risks involved are managed,
- have the right people doing the right work at the right time,
- cooperate and coordinate work,
- have the right information about risks and how they are managed,
- effectively communicate information to those who need it, and
- consult and engage workers about risks and how they are managed.

The purpose of the CDM Regulations is to enable those involved with construction to manage the health, safety and welfare of those affected whilst construction work takes place.

Regulation 2 sets out how the Regulations are to be interpreted and describes what is meant by construction work, and lists the role of persons (dutyholders) involved in construction work such as; client, designer and contractor for example. Table 1 highlights the dutyholders and briefly outlines their roles and responsibilities and is an extract from the HSE guidance to CDM Regulations.

Table 1 Roles and responsibilities of dutyholders

CDM duty holders	Who are they?	Summary of role / main duties
Clients	Organisations or individuals for whom a construction project is carried out.	Make suitable arrangements for managing a project. This includes making sure that other duty holders are appointed and sufficient time and resources are allocated. Relevant information is prepared and provided to other duty holders.
Domestic clients	People who have construction work carried out on their own home or the home of a family member, that is not done in furtherance of a business, whether for profit or not.	Domestic clients are in scope of CDM 2015 but their duties as a client are normally transferred to the contractor, on a single contractor project or the principal contractor, on a project involving more than one contractor.
Principal designers	Are appointed by the client in projects involving more than one contractor. They can be an organisation or an individual with sufficient knowledge, experience and ability to carry out the role.	Plan, manage, monitor and co-ordinate health and safety in the pre-construction phase of a project. This includes amongst other duties preparing and providing relevant information to other duty holders.
Designers	Those who, as part of a business, prepare or modify designs for a building or product, or prepare or modify designs to systems relating to construction work	Provide information to other members of the project team to help them fulfil their duties.
Principal contractors	Appointed by the client to co-ordinate the construction phase of a project where it involves more than one contractor.	Plan, manage, monitor and co-ordinate health and safety in the construction phase of a project.
Contractors	Those who do the actual construction work. They can be either an individual or a company.	Plan, manage and monitor construction work under their control so that it is carried out without risks to health and safety. For projects involving more than one contractor, co-ordinate their activities with others in the project team. For single-contractor projects, prepare a construction phase plan.
Workers	The people who work for or under the control of contractors on a construction site.	Responsible for their own health and safety and that of others who may be affected by their actions.

Regulation 4 states that it is the duty of the client to make suitable arrangements for the management of the project. It is the client's responsibility to make sure that before the construction phase begins, the contractor draws up a construction phase plan and the designer prepares a safety file which is to be revised appropriately as the project continues (Regulations 12 and 15 refer).

Regulation 6 states that a project is notifiable if the construction work on a construction site is scheduled to -

(a) last longer than 30 working days and have more than 20 workers working simultaneously at any point in the project; or

(b) exceed 500 person days.

Where a project is notifiable, the client must give notice in writing to the Health and Safety Executive (HSE) as soon as is practicable before the construction phase begins. The notice must comply with the requirements of Regulation 6(3).

Regulation 8 recognises that organisations or individuals can carry out the role of more than one dutyholder, provided they have the skills, knowledge, experience and (if an organisation) the organisational capability necessary to carry out those roles in a way that secures health and safety of all concerned involved with the project.

CDM 2015 Regulations makes a distinction between domestic and commercial projects. Client duties apply in full for commercial projects where the client has the overall responsibility to see that the project is fully managed. For domestic projects, the client's duty normally passes to other dutyholders such as the principal contractor.

1.2.4 Provision and Use of Work Equipment Regulations 1998

The Provision and Use of Work Equipment Regulations require that work equipment, including installations, is so constructed (or adapted) so as to be suitable for the purpose for which it is provided.

Equipment must be inspected after installation and before use, at suitable intervals and after exceptional events for example fire, flooding, or mechanical damage. Potentially dangerous machinery must be guarded. Where necessary, equipment logs must be maintained and adequate training given to equipment operators.

1.2.5 Electricity Safety, Quality and Continuity Regulations

The main purpose of the Electricity Safety, Quality and Continuity Regulations is to provide for the safety of the electricity supply distribution system, that is, for the safety of the general public and of persons working on the system. As the title implies, the Regulations also have requirements for power quality and supply continuity. There are many references to *BS 7671*. The regulations require distributors, prior to connection of new properties, to obtain confirmation that the installation to be connected complies with the requirements of *BS 7671* (Regulations 9(4) and 25(2) refer).

Regulation 24(4) requires the distributor to make available an earth connection for new supplies, unless inappropriate for reasons of safety. In the UK this is almost always achieved by the provision of a PME earthing facility.

Regulation 27(3) recognises that, depending upon the distance from the substation transformer, the characteristics of the distribution network and the loading on the system, the voltage and the frequency of the supply at the consumers' supply terminals will vary. In the case of a low voltage supply clause 3 (b) limits the permitted variation in voltage at the consumers' supply terminals to plus 10 % and minus 6 % of the declared voltage.

Regulation 28 requires the distributor to provide on request a written statement of:

a) the maximum prospective short-circuit current at the supply terminals;

b) for low voltage connections, the maximum earth loop impedance of the earth fault path outside the installation;

c) the type and rating of the distributor's protective device or devices nearest to the supply terminals;

d) the type of earthing system applicable to the connection; and

e) the number of phases, frequency and voltage.

1.2.6 The Building Regulations

England and Wales. The Building Regulations 2010

The Building Regulations 2010 (Statutory Instrument 2010 No. 2214) are made under powers provided in the Building Act 1984 and currently apply in England and Wales, but not Scotland or Northern Ireland.

The purpose of the Building Regulations is to provide for the health and safety of people in and around buildings and also provide for matters such as energy conservation and access and use. In the Building Regulations the term "building work" means:

a) the erection or extension of a building,

b) the provision or extension of a controlled service or fitting in or in connection with a building,

c) the material alteration of a building, or a controlled service or fitting,

d) work required by Regulation 6 (requirements relating to material change of use),

e) the insertion of insulating material into the cavity wall of a building,

f) work involving the underpinning of a building.

Schedule 1 of the Building Regulations describes the requirements relating to building work and material change of use as described in Regulations 4 & 6. These requirements are divided into the following parts:

Part A	(Structural safety)
Part B	(Fire safety)
Part C	(Resistance to contaminants and moisture)
Part D	(Toxic substances)
Part E	(Resistance to the passage of sound)
Part F	(Ventilation)
Part G	(Sanitation, hot water safety and water efficiency)
Part H	(Drainage and waste disposal)
Part J	(Combustion appliances and fuel storage systems)
Part K	(Protection from falling, collision and impact)
Part L	(Conservation of fuel and power)
Part M	(Access to and use of buildings)
Part N	(Glazing safety)
Part P	(Electrical safety)
Part Q	(Security)
Part R	(Physical infrastructure for high speed electronic communications networks)

To provide the construction industry with guidance on how these requirements of the Building Regulations can be met, for each of the parts previously listed, the Ministry of Housing, Communities and Local Government publishes a series of Approved Documents.

The Approved Documents may be downloaded free-of-charge from **www.planningportal.gov.uk**
Two editions of Approved Document P, dealing with electrical safety, are available:

- the 2006 Edition,
- the 2013 Edition, which applies to work in England from 6 April 2013.

It should be noted that the power to make Building Regulations for Wales has been transferred to Welsh Ministers from 31st December 2011. The Building Regulations 2010 and related guidance such as the Approved Documents published prior to that date will continue to apply in Wales until Welsh Ministers decide that changes should be made. As a result of this transfer of power, for electrical installations in dwellings, the approved document used in England is different to that used in Wales. In England, the 2013 edition of Approved Document P is used whilst, in Wales, apart from 'excepted energy buildings[1]', the 2006 (including 2010 amendments) edition is used.

[1]Excepted energy buildings, as defined in The Welsh Ministers (Transfer of Functions) (No. 2) Order 2009, are energy infrastructure buildings and, for these buildings, the 2013 edition of Approved Document P is applicable.

Although there are significant differences between the Approved Documents used in England and Wales both approved documents:

1. Limit the scope of Part P to electrical installations in dwelling-houses, flats and certain common areas and business premises associated with the dwelling house or flat.
2. Contain statements to the effect that electrical installation work should be designed and installed in accordance with *BS 7671*. Both editions of Approved Document P require all electrical installation work to be inspected and certified in accordance with the requirements of *BS 7671*.

However, in terms of the types of work that is classified as being 'notifiable electrical installation work' and the category of person permitted to inspect and test notifiable work there are important differences between the approved documents used in England and Wales. In Approved Document P the term 'notifiable electrical installation work' means electrical installation work that must be either certified by the building control body or certified by a person authorised by the applicable approved document. After completion of the notifiable electrical installation work, whichever procedure is followed, the occupier of the building should receive a copy of a Completion Certificate.

Notifiable work - Wales

Section 0 of Approved Document P: 2006 provides details of notifiable work and these details are summarised below. The requirements for work in kitchens, special locations and special installations are different from the requirements for the general areas of the dwelling.

Notes:

1. A kitchen (as defined in Section 0 of Approved Document P: 2006 is a location containing a sink and food preparation facilities).
2. Special locations, as given in Table 2 of Approved Document: 2006, are locations containing a bath tub, shower basin, swimming pool, paddling pool and a hot air sauna.
3. Special installations are installations that, due to the nature of the installation, additional dangers exist. These are described in Table 2 of Approved Document: 2006 and are described in the following section.

The electrical installation work that is currently notifiable in Wales is as follows:

Anywhere within or attached to a dwelling:

- a complete new installation (a full rewire),
- replacing a consumer unit,
- the installation of:
  - > a new circuit,
  - > the power or control wiring for a new central heating system,
  - > a special installation as given in Table 2 of Approved Document: 2006 - are electric floor and ceiling heating systems, garden lighting and power installations, a solar photovoltaic power supply system, a small scale generator or an extra-low voltage lighting installations, (other than a pre-assembled, CE marked lighting sets).

Within a kitchen or special location:

The work described in the previous section, plus:

- any alteration or addition to an existing circuit for the addition of adding lighting points, socket-outlets or a fused-spur,
- the supply to another location, such as a conservatory and attached garages.

Within a special location

- Installation of telephone or extra-low voltage wiring and equipment for the purposes of communications, information technology, signalling, control or similar.

Notifiable work - England

When compared with the description of the electrical installation work that is notifiable in Wales (based on the 2006 edition of Approved Document P), the list of the electrical installation work that is currently notifiable in England (based upon the 2013 edition of Approved Document P) is, as shown in the following section, less comprehensive.

The electrical installation work that is currently notifiable in England is as follows:

Anywhere in a dwelling or its surroundings:

- installation of a new circuit,
- replacement of a consumer unit.

Within a special location:

- any alteration or addition to an existing circuit.

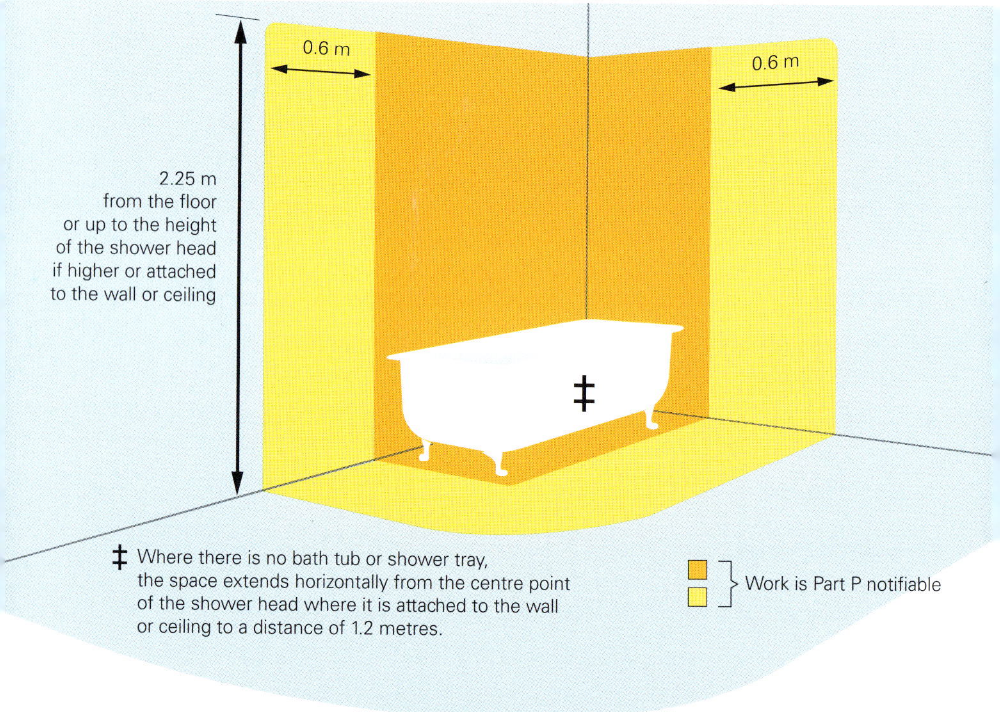

Fig 1.1 Illustration for notifiable work in a room containing a bath or shower in accordance with Approved Document P: 2013

As described in Approved Document P: 2013, special locations in dwellings are rooms containing a bath or shower, swimming pool or a sauna heater. For a room containing a bath or shower, an alteration or addition to an existing circuit is notifiable only where carried out in the space surrounding a bath or shower (as illustrated in Fig 1.1), rather than the whole room volume, which is the case for rooms containing a swimming pool or sauna heater.

Certification of notifiable work

The 2006 edition of Approved Document P permits the certification of notifiable work to be carried out by either:

1. an installer who is registered with a Part P competent person self-certification scheme such as the ones offered by NICEIC or ELECSA, or

2. a building control body.

In addition to the two methods of inspection and testing of notifiable work permitted by the 2006 edition of Approved Document P, the 2013 edition of this document permits certification by a registered third party certifier. However, due to concerns regarding the safety of the installation, operators of many self-certification schemes, including NICEIC and ELECSA, do not operate a 3rd party certification scheme.

The 2013 edition of Approved Document P advises that to comply with the Building Regulations, depending upon the nature of the work being undertaken, it may be necessary to comply with other approved documents and provides the following examples:

Part A (Structural): depth of chases in walls, and the size of notches in floors and roof joists

Part B (Fire safety): Fire safety of certain electrical installations; provision of fire alarms and fire detection systems; fire resistance of service penetration through floors, walls and ceilings

Part C (Resistance to contaminants and moisture): resistance of service penetrations to rainwater and contaminants such as radon

Part E (Resistance to the passage of sound): soundproofing of service penetration

Part F (Ventilation): dwelling ventilation rates

Part L (Combustion appliances and fuel storage systems): energy efficient lighting

Part M (Access to and use of buildings): height of socket-outlets and switches

Further information on how the Building Regulations affect the work of an installation electrician can be found in the in the NICEIC and ELECSA publication 'A Short Guide to the Building Regulations'. Within this Guide advice on how to comply with Part A (Structure) is provided in Section 11.6.

The responsibility for complying with the Building Regulations rests with the person carrying out the building work, which can include plumbing and electrical installations. The person carrying out the electrical installation is personally responsible for compliance. Persons employing others to carry out work should confirm who has responsibility for compliance with the Building Regulations.

The owner of the building may be served with the enforcement notice if the work does not comply with the Building Regulations.

Scotland. Building (Scotland) Regulations

The Building (Scotland) Regulations are made under powers provided in the Building (Scotland) Act 2003. This Act gives Scottish Ministers the power to make building regulations to provide for the health, safety, welfare and convenience of persons using buildings and of those who may be affected by buildings or matters connected with buildings such as how they are used by others. The Building Regulations also seek to promote the conservation of fuel and power and encourage sustainable development.

Two Technical Handbooks have been issued by Scottish Ministers to provide guidance on how to comply with the standards set in the Building (Scotland) Regulations; one for domestic buildings and one for non-domestic buildings.

The handbooks consist of the following sections:

0	General	4	Safety
1	Structure	5	Noise
2	Fire	6	Energy
3	Environment	7	Sustainability

It is worth noting at this point that a dwelling may include surgeries, consulting rooms and the like having a floor area not exceeding 50 m^2 forming a part of a dwelling and which are used by the occupier of the dwelling for professional or business purposes (Section 0.2.2 of the Technical Handbook - Domestic refers).

Where a building has mixed domestic and non-domestic usage, the relevant part(s) of the appropriate technical handbook should be applied. Where a particular building or part thereof might be considered to fall into more than one category, the building should be designed to meet the more stringent of the recommendations (see Section 0.1.6. of the Technical Handbook - Domestic).

The handbooks and associated appendices may be downloaded, free-of-charge, from the Scottish Government website **www.gov.scot**

1.2.7 BS 7671

BS 7671: 2018 ***Requirements for Electrical Installations***, also known as the *IET Wiring Regulations* **Eighteenth Edition,** contains requirements for the design, erection and verification of most types of general electrical installation (Regulation 110.1 refers).

Given the ever increasing complexity of electrical installations, it is frequently necessary to also take into account the requirements and recommendations contained within other British Standards, Codes of Practice and the like.

Certain electrical installations of a more specialised nature, such as those in aircraft, ships and trains; public systems for electrical distribution; and within a general installation, including lightning protection systems, lifts and the electrical equipment of machines, are not covered by *BS 7671* but rather by other British Standards (Regulation 110.2 refers).

BS 7671: 2018 introduced a new informative appendix where the objective is to consider how energy usage can be reduced.It is beyond the scope of this Guide to consider further how the application of Appendix 17 impacts upon the design and erection of electrical installations. whilst maintaining electrical safety and control.

Although the requirements of *BS 7671* are non-statutory, compliance with them may be cited as evidence in a court of law to claim compliance with a statutory requirement (see Regulation 114.1). The Health and Safety Executive (HSE) states in a note within the introduction of *BS 7671* that they (HSE) consider installations which conform with the relevant parts of *BS 7671* as being likely to achieve conformity with the relevant parts of the *Electricity at Work Regulations 1989.*

1.3 Guidance

1.3.1 Health and Safety Executive HSR25 - The Electricity at Work Regulations 1989 - Guidance on Regulations

This Health and Safety (HSE) publication provides clear, concise guidance on the general regulations contained within the Electricity at Work Regulations 1989 and is essential reading for any persons involved in electrical installation work in the UK. It is now downloadable – free of charge – from the HSE website.

Anyone who is involved with electricity, whether working directly on electrical systems, installations or equipment, or using electrically operated equipment, is classed as a Duty Holder. As has already been stated in Section 1.2.2 of this Guide, Regulation 3 of EWR requires duty holders to comply with the provisions contained within EWR in so far as they relate to matters within their control.

Depending on the particular regulation concerned, the regulations in EWR 1989 impose one of two levels of duty upon persons working on, or in close proximity to, electrical systems (as defined in Regulation 2 of EWR 1989) or parts thereof, namely:

- absolute, or
- so far as is reasonably practicable.

Absolute duty

An absolute duty requires that a regulation must be met regardless of the cost, time or effort incurred in so doing. Regulation 29 offers a defence where criminal proceedings have been brought against a person or persons considered to have failed in their absolute duty. The defence puts the onus on the person(s) so accused to demonstrate that 'they took all reasonable steps and applied due diligence to avoid' committing an offence.

Unless stated otherwise, the requirements of a particular regulation are absolute.

Reasonable practicability

Where embodied in a regulation, reasonable practicability allows a balance to be struck between, on the one hand, the level of risk reduction considered acceptable and, on the other, the cost, time and effort of so doing.

In order to apply reasonable practicability, it will be necessary for a risk assessment to be performed and it should be remembered that this risk assessment could be the subject of scrutiny in the event of an accident or incident occurring.

Competence

Regulation 16 requires persons to be competent to prevent danger and injury. It should be noted that this competence requirement applies not only to those persons involved in work activities on or near electrical installations and equipment, but also to those employed in a managerial, design or supervisory capacity whose actions and decisions might put themselves or, more likely, others at risk of danger or injury.

HSR25 makes it clear that the intention of Regulation 16 is 'to ensure that persons are not placed at risk due to a lack of skills on the part of themselves or others' when dealing with electrical equipment.

Regulation 16 cites technical knowledge and experience as being appropriate indicators of competence and continues to state that in some cases an appropriate degree of supervision may be necessary. Although the successful completion of a particular qualification does not of itself indicate an individual's competence, it might act as a suitable indicator of such.

Similarly, having a number of years of relevant work experience does not mean that a person is competent, but it might do so.

A person may be deemed competent at a particular work activity or range of work activities on account of the relevant training and experience that has been accumulated over time. Where a company employs qualified and experienced personnel, there may be occasions when unfamiliar equipment and / or working practices could be encountered that might require further training to be provided.

Supervision by a competent person can be an appropriate means of allowing those who are not yet fully competent (such as trainees) to gain valuable work experience in a safe manner. It can also permit persons who are not competent, such as painters decorating an electrical switchroom, to work in close proximity to electrical equipment without risk of danger or injury. In such circumstances *BS 7671* would consider them to be instructed persons (as defined).

A person carrying out inspection and testing activities must be competent in such work in terms of their experience and training, and where necessary any shortfall in either of these factors may be addressed by the provision of appropriate supervision.

In Northern Ireland the Electricity at Work Regulations (Northern Ireland) 1991 apply. The Health and Safety Executive for Northern Ireland (HSENI) has produced the *Memorandum of Guidance on the Electricity at Work Regulations (Northern Ireland) 1991* to give guidance on these Regulations.

GS38 - Electrical test equipment for low voltage electrical systems

This HSE publication gives guidance on the following in connection with the selection and use of voltage test lamps and voltage detectors:

- causes of accidents,
- requirements for test probes and leads,
- systems of work,
- examination of equipment.

Precautions to be taken during testing

The key recommendations of GS38 are as follows.

i) Test probes should have finger barriers to prevent hands inadvertently slipping and allowing accidental contact with live parts.

ii) The tips of the probes should be insulated such that no more than 4 mm of the tip remains exposed.

iii) Wherever practicable – that is, where this will not interfere with the correct use of the test lamp or voltage detector, no more than 2 mm of tip should be exposed, or probes fitted with spring loaded insulated screening should be used.

iv) Leads should be adequately insulated and identified.

v) Leads should also be sufficiently flexible and of length suitable for their purpose, that is, neither too long nor too short.

vi) Where the leads are not permanently attached to the instrument, there should be no accessible live parts in the event of a lead becoming detached in use.

vii) In the case of voltage test lamps, the leads (and therefore the person using the instrument) should be protected by either:

- a high breaking capacity (HBC) fuse of rating not exceeding 500 mA, or
- a current limiting resistor and a fuse.

viii) In the case of voltage detectors, the current that may pass through the instrument is limited by the internal circuit design, protective impedance and current limiting resistors built into the probes.

ix) The maximum voltage and, where applicable, any short time rating should be clearly shown on voltage test lamps and voltage detectors.

Guidance Note GS38 is available as a free download from: **www.hse.gov.uk**

Other Guidance

The HSE publish guidance on a wide range of topics which are of relevance to those involved in electrical installation work including the following.

- Simple guide to the Provision and Use of Work Equipment Regulations 1998
- Safe use of work equipment. Approved Code of Practice and Guidance (L22)
- Short guide to the Personal Protective Equipment at Work Regulations 1992 (L25).

1.3.2 Electrical Safety First

Electrical Safety First formerly known as the Electrical Safety Council produces, in association with other industry bodies, a range of technical and best practice guidance for the electrical industry, as well as information on electrical product safety. Of particular interest to installation electricians are the Best Practice Guides and the questions and answers produced by the Electrical Installation Forum.

Best Practice Guides

There are currently nine Best Practice Guides:

No 1 Replacing a consumer unit in domestic premises, where lighting circuits have no protective conductor

No 2 Guidance on the management of electrical safety and safe isolation procedures for low voltage installations

No 3 Connecting a micro-generation system to a domestic or similar electrical installation (in parallel with the mains supply)

No 4 Electrical installation condition reporting: Classification Codes for domestic and similar electrical installations

No 5 Electrical installations and their impact on the fire performance of buildings

Part 1 - Domestic premises: Single family units (houses, flats, maisonettes, bungalows)

No 6 Portable and stationery appliance testing in private rented accommodation

No 7 Test instruments for electrical installations: Accuracy and consistency

No 8 Selection and use of plug-in socket-outlet test devices.

No 9 Safe installation of retrofit LED lamps

All of the above are freely available from the ESF website on the installation forum

www.electricalsafetyfirst.org.uk

1.3.3 NICEIC and ELECSA

NICEIC and ELECSA have produced a wide range of publication relating to electrical installation work. Included in these publications are the popular Short Guide series which deal with specific topics, a number of Learning Guides designed to assist electricians further develop their technical skills, and Snags and Solutions which provide guidance on practical problems which may occur on site and also guides to Initial and Periodic Verification.

Details of NICEIC and ELECSA publications can be found in the electrical section of NICEIC Publications at the NICEIC shop. **www.shop.niceic.com**

The Wiring Regulations Advisory Group (WRAG)

WRAG is a discussion forum comprising of technical representatives from a wide range of industry bodies, hosted by the Electrical Safety First. It produces guidance for electrically competent persons on a range of technical issues that have been raised in enquiries to the bodies contributing to the forum.

Topics relating to new work, additions and alterations, certification and reporting, and periodic inspection and testing have been discussed with the resulting guidance made freely available from the Electrical Safety First website **www.electricalsafetyfirst.org.uk**

Other Guidance

Electrical Safety First, with the co-operation and assistance of relevant industry partners, has produced guidance on matters such as downlighters and fire safety, electrical safety for landlords and electrical safety for those involved in the construction industry. This guidance is also downloadable, free-of-charge, from the Electrical Safety First website.

CHAPTER 2
SUPPLY SYSTEMS AND EARTHING & BONDING

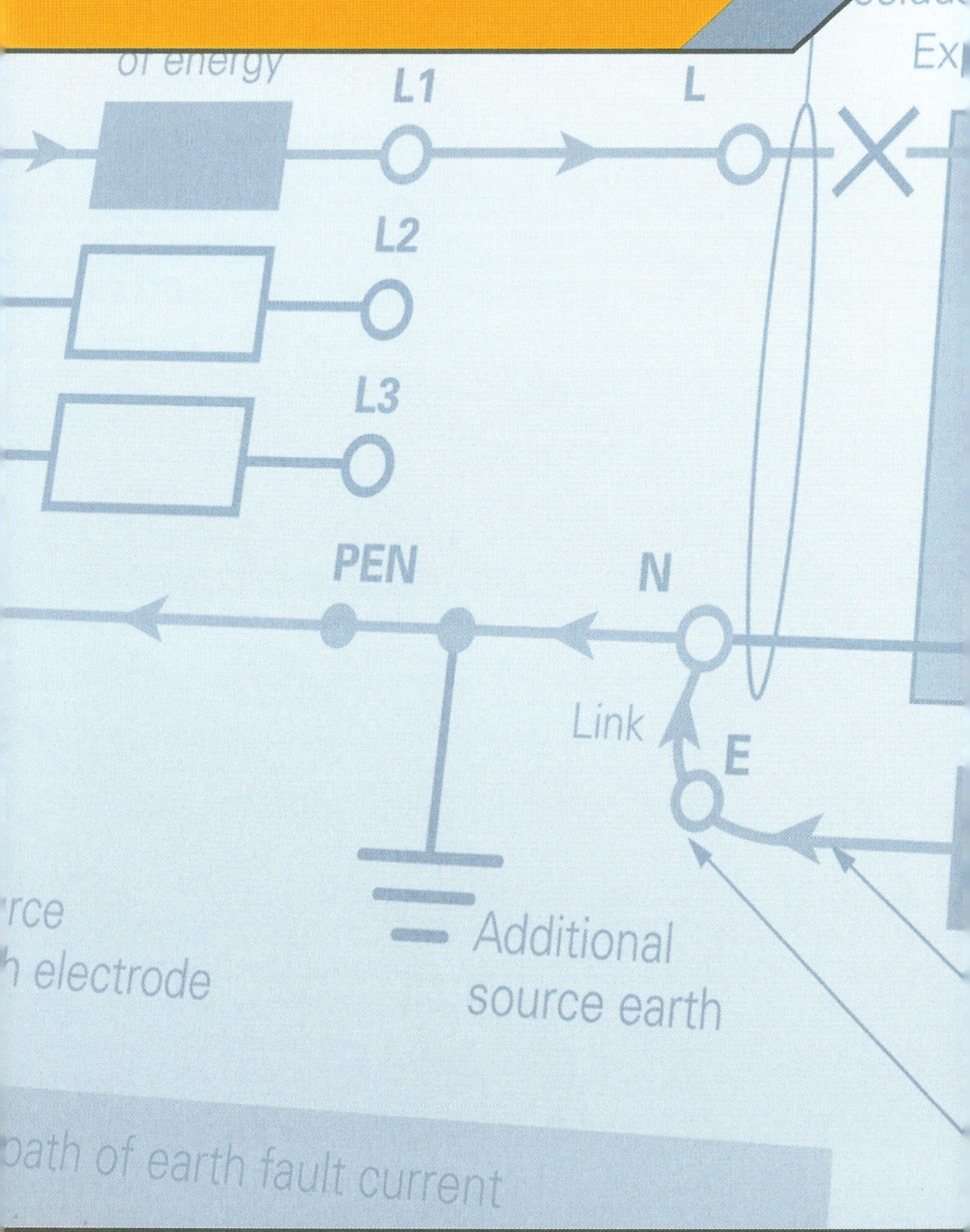

2.1 The service position

The consumer's electrical installation begins at the outgoing terminals of the electricity supplier's meter. All equipment upstream of these terminals, including the meter itself, typically is the property of the meter operator, electricity supplier or electricity distributor.

The electricity supply to the dwelling is subject to the statutory requirements of the *Electricity Safety, Quality and Continuity Regulations*.

Note. Further information regarding systems and their earthing arrangements may be obtained from the electricity distributors via the Energy Networks Association **www.energynetworks.org**

2.2 System types and earthing arrangements

Within the scope of this Guide, the electrical system, which comprises the electricity supply and the whole of the consumer's electrical installation, will be one of three distinct types: TN-S, TN-C-S or TT. The system type will determine the earthing arrangements for the installation and influence its design.

TN-S, TN-C-S and TT systems each have different regulatory requirements in respect of protection against electric shock, protection against overcurrent, and isolation and switching. It is therefore essential in terms of safety that the electrical designer and electrical contractor responsible for an installation know with certainty the type of system of which the installation forms part. For this reason, Regulation Group 312.2 of *BS 7671* requires the type of system earthing to be assessed.

The three types of earthing system are distinguished from one another by the specific methods used to earth the source of supply and the installation, and also by whether the neutral and protective functions of the system are performed by separate conductors, a combined conductor, or a mixture of the two (in different parts of the system).

For all three system types mentioned above:

- at least one point of the source of supply is directly connected to Earth (for example the neutral of a single-phase transformer or the star point of a three-phase transformer),
- the neutral and protective conductors of the installation are separate throughout the installation.

For a TN-S system, the neutral conductor and the protective conductor are separate in the supply (see Fig 2.1).

[1]Electricity distributors sometimes use the term 'CNE' (Combined Neutral Earth) instead of 'PEN', for a combined protective and neutral conductor

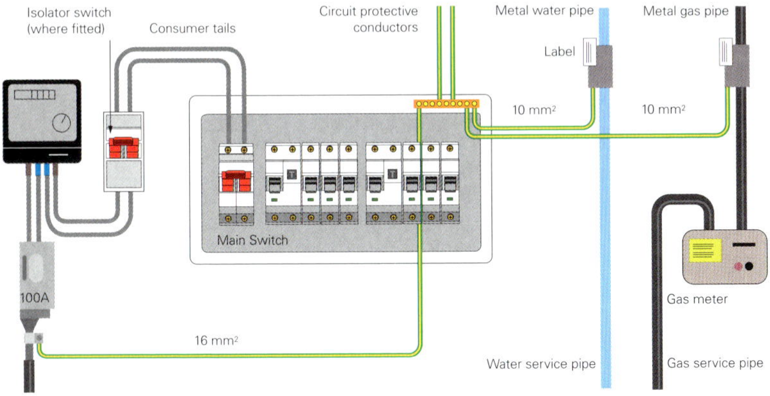

Fig 2.1 Typical TN-S earthing and bonding arrangement

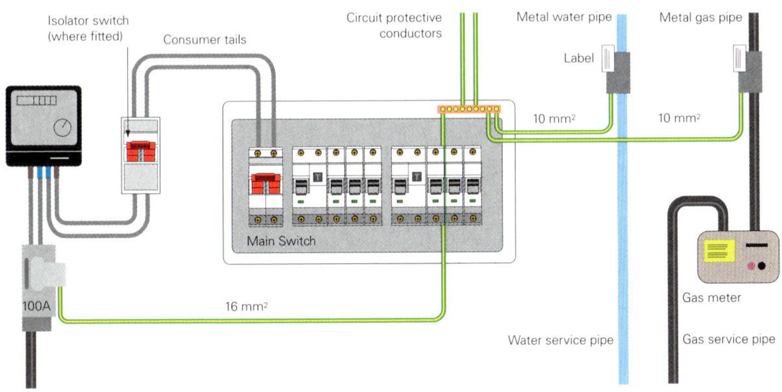

Fig 2.2 Typical TN-C-S earthing and bonding arrangement

For a TN-C-S system, the neutral conductor and the protective conductor are combined into one conductor in the supply usually called a PEN[1] conductor (see Fig 2.2).

For a TT system, no protective conductor is available with the supply, or if a protective conductor is available, it is not used for the electrical installation (as might be the case for some special installations or locations covered within Part 7 of *BS 7671*), and the installation is earthed by means of an installation earth electrode (see Fig 2.3).

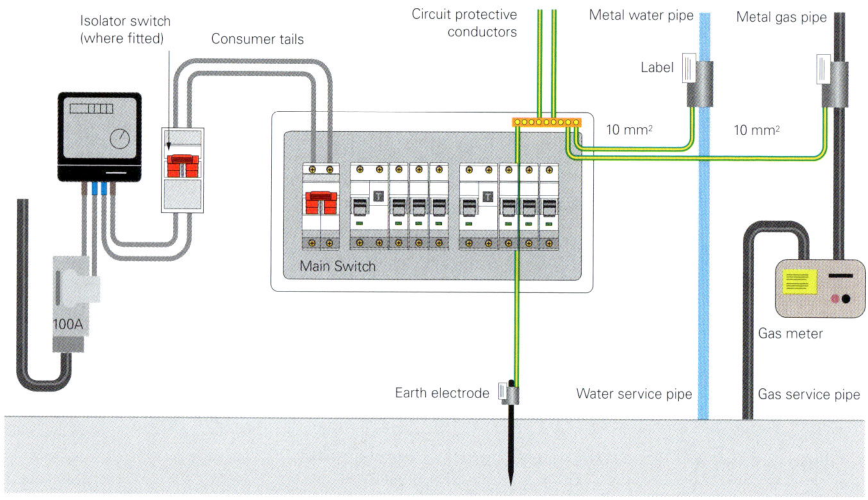

Fig 2.3 Typical TT earthing and bonding arrangement

TT systems are often employed where the electricity distributor is unable on technical or safety grounds to offer an earthing facility for the installation, such as for premises in areas of low population density where it is not economically viable for the network to carry out multiple earthing.Table 2.1 summarises the earthing arrangements for the three types of system.

Note: See Fig 2.7 (a)

Table 2.1 System types and earthing arrangements

System type	Supply		Installation		Available means of earthing
	Supply neutral conductor[1]	Supply protective conductor[1]	Installation neutral conductor[1]	Installation protective conductor[1]	
TN-S	S	S	S	S	Distributor's earthing terminal
TN-C-S	C	C	S	S	Distributor's earthing terminal
TT	S	N/A	S	S	Consumer's installation earth electrode

1: S=Separate, C = Combined, = Not Available

2.3 Supply system earthing

Irrespective of system type (TN-S, TN-C-S or TT), the supply to an electrical installation is earthed at its source of energy, for the purpose of safety and for referencing supply voltages.

Normally, the neutral (or star) point of the source transformer secondary winding is connected to the general mass of Earth, via a source earth electrode. The provision and maintenance of source earthing is normally the responsibility of the owner of the source of energy, usually the electricity distributor.

2.4 Installation earthing

Electrical installation earthing is the connection of the exposed-conductive-parts of an electrical installation to an appropriate means of earthing. It is applicable in installations where protection against electric shock is provided by Automatic Disconnection of Supply (ADS), as is the case in the majority of domestic installations in the United Kingdom.

The purpose of electrical installation earthing is to allow earth fault current to flow in order to operate the protective device for protection against electric shock, thereby providing automatic disconnection.

The principal components of the earthing system of an electrical installation are shown diagrammatically in Figs 2.4 and 2.5 and include:

- circuit protective conductors,
- Main earthing terminal (MET),
- earthing conductor of the electrical installation, and
- means of earthing.

The most commonly-used types of electrical system, a TN-C-S system with Protective Multiple Earthing (PME) applied is shown in Fig 2.4.

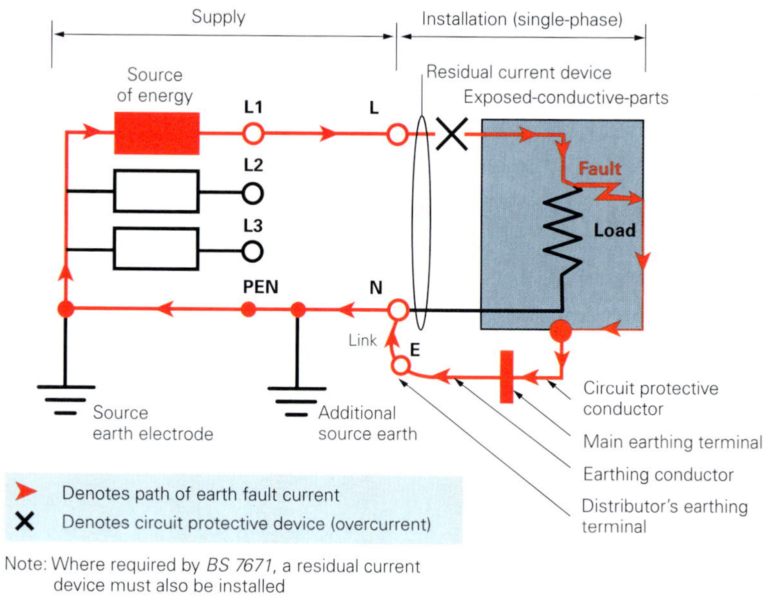

Fig 2.4 Electrical equipment earthing showing the earth fault current path for a typical TN-C-S system with Protective Multiple Earthing (PME) applied.

The installation shown in Fig 2.4 has an earth fault, the path of the fault current being represented by the red arrows. The loop impedance of this path should not exceed the maximum ohmic value permitted by Section 411 of *BS 7671*, according to the type and rating of the protective device installed to provide automatic disconnection of the circuit in question under earth fault conditions.

Provided this loop impedance requirement is met, then automatic disconnection can be expected to occur within the maximum time permitted by Chapter 41 in the event of an earth fault of negligible impedance.

For electrical installations forming parts of other types of system, different means of earthing are applicable. For example, the appropriate means of earthing for an installation forming part of a TT system is an installation earth electrode, as shown in Fig 2.5.

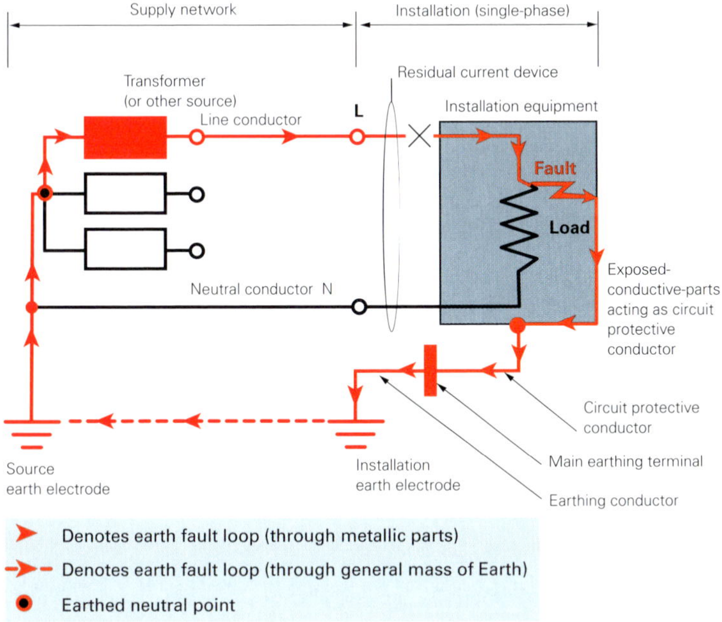

Fig 2.5 Electrical equipment earthing showing the earth fault current path for a typical TT system.

In a TT system, the path of the earth fault current includes the source earth electrode.

2.5 Means of earthing

A means of earthing is an arrangement through which a connection is made to the general mass of Earth for the exposed-conductive-parts of an installation. The exposed-conductive-parts are connected to the means of earthing via the earthing conductor and main earthing terminal (MET) of the installation (see Fig 2.6).

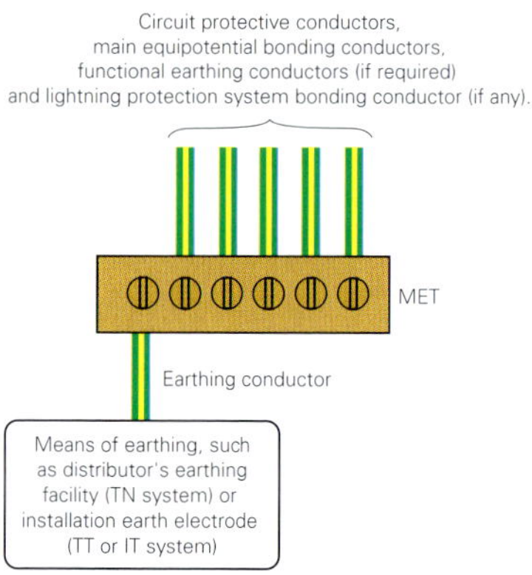

Fig 2.6 Means of earthing for an installation in relation to the earthing conductor and main earthing terminal.

The particular means of earthing for an installation depends largely on the type of system of which the installation forms part. For an installation forming part of a TT system, the means of earthing is an installation earth electrode. For an installation connected to an electricity distributor's network and forming part of a TN-S or TN-C-S system, the means of earthing is normally provided by the electricity distributor. In such a case, the particular means will depend upon the composition of the distributor's supply lines (Regulations 542.1.2.1 to 542.1.2.3 refer).

The main features of each of the different means of earthing are detailed in Table 2.2, including references to the use of facilities provided by the electricity distributor where appropriate.

Table 2.2 System types and earthing arrangements

System type	Means of earthing	Remarks (main features)
TN-S	Connection of the MET of the installation to the earthed point of the source of energy via a separate protective conductor. Where a distributor provides the low voltage supply, this connection is via the distributor's supply lines, normally using: • the metallic sheath or armour of the supply cable, or • the protective conductor of a split - concentric supply cable, or • a separate protective conductor (for example, of an overhead line)	Connection to a protective conductor of a distributor's supply cable or overhead line may be carried out only with the express permission of, and by a method approved by, the distributor. Many distributors will provide, on request, an earthing terminal connected to the protective conductor of their supply cable or overhead line. The MET of the electrical installation is to be connected to this terminal by means of the earthing conductor.
TN-C-S (for example where protective multiple earthing [PME] is provided)	Connection of the MET of the installation to the earthed neutral of the source of energy via the combined neutral and earth PEN conductor of the distributor's low voltage supply lines, normally using: • the PEN conductor of the supply cable, or • the PEN conductor of the overhead line.	The distributor provides an earthing terminal connected to the PEN conductor of the supply cable or overhead line. The MET of the installation is to be connected to this terminal via the earthing conductor.
TT	Connection of the MET to the installation earth electrode.	The installation earth electrode forms part of the installation.

2.6 Responsibility for providing, assessing and verifying the means of earthing

In most instances electricity distributors will make available a means of earthing for an electrical installation. However, ensuring an installation is adequately connected with Earth is always the responsibility of the installation owner, who usually delegates the task to an electrical contractor. For first-time supplies, the electricity distributor is obliged under the *Electricity Safety, Quality and Continuity Regulations* to make available an earthing facility for the consumer's installation, unless there are reasons of safety which preclude such a provision.

The designer of an electrical installation, or of any alteration or addition to an installation, is responsible for assessing the intended means of earthing, by inspection, testing or, where appropriate, enquiry to the distributor, to ensure that it is suitable for the needs of the installation in terms of safety (Regulations 132.16 and 313.1 refer).

It should never be assumed that what may seem to be an existing earthing facility, such as an earthing terminal apparently connected to the supply cable sheath or supply neutral, is an effective means of earthing. Appropriate verification must always be carried out.

2.7 Main protective bonding

Protective bonding is an indispensable part of the most commonly used protective measure against electric shock; Automatic Disconnection of Supply (ADS). Typically, such bonding is required in the majority of electrical installations where one or more circuits rely on ADS.

An extraneous-conductive-part is defined in *BS 7671* as:

> A conductive part liable to introduce a potential, generally Earth potential, and not forming part of the electrical installation.

Regulation 411.3.1.2 requires that, in each installation, protective bonding conductors complying with Chapter 54 should connect with the main earthing terminal, all extraneous-conductive-parts of that installation, including:

- water installation pipes,
- gas installation pipes,
- other installation pipework and ducting,
- central heating and air conditioning systems,
- exposed metallic structural parts of the building.

It should be noted, that the items listed above do not always fall within the definition of extraneous-conductive-part, nor is the list exhaustive.

Where appropriate, it will be necessary to connect a lightning protection system to the protective equipotential bonding in accordance with *BS EN 62305*.

BS 7671 also requires main protective bonding to be connected to any metallic sheath of a telecommunication cable. However, the consent of the owner or operator of the cable must be obtained before so doing.

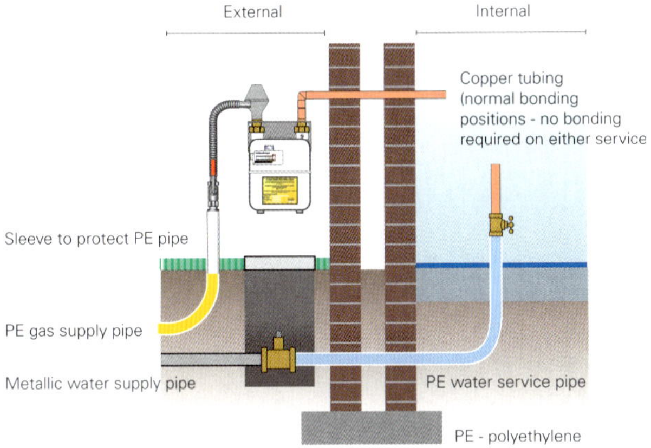

Fig 2.7 (a) Water and gas service (installation) pipework run in non-metallic material

Where any metallic pipes enter a building have an insulating section at their point of entry, as shown by Fig 2.7(a), there is no requirement to provide protective equipotential bonding (Regulation 411.3.1.2 refers).

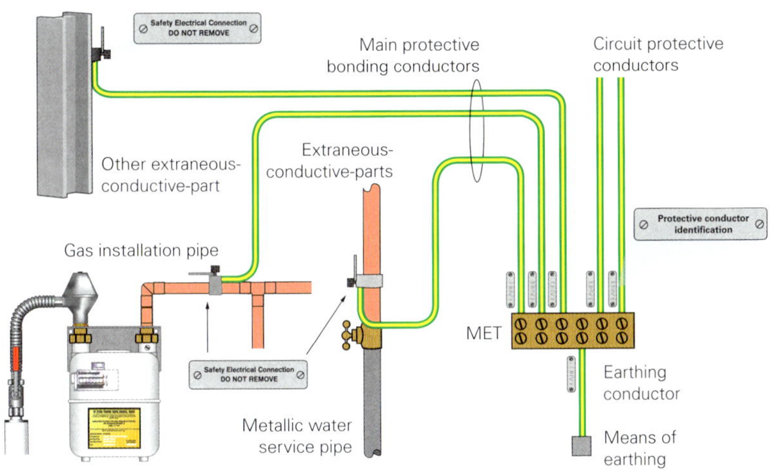

Fig 2.7 (b) Protective equipotential bonding requirements.

Where an installation has metallic gas and water services, these are extraneous-conductive-parts and therefore, typically require protective equipotential bonding to be carried out as shown in Fig 2.7(b).

Where an installation serves more than one building, such as a house with a detached garage, the above requirements apply to each building.

Protective bonding conductors in PME-supplied installations may have to carry sustained currents under certain distribution network conditions, which may raise the temperature of the bonding conductors. Where PME conditions apply, the electricity distributor may have particular requirements for protective bonding conductors, above and beyond those of *BS 7671*.

The requirements for the minimum permitted cross-sectional area (csa) of a main protective bonding conductor are given in Regulation 544.1. Protective bonding conductors that are buried in the ground are, additionally, subject to the requirements of Regulation 542.3.1 for earthing conductors.

Where PME conditions **do not apply**, Regulation 544.1.1 requires a main protective bonding conductor to have a csa of not less than half that required for the earthing conductor of the installation and not less than 6 mm^2. The csa need not exceed 25 mm^2 if the main bonding conductor is of copper or, if of other metals, a csa affording equivalent conductance.

Table 2.3 Minimum protective bonding conductor csa in relation to the PEN conductor where PME conditions apply

Csa of PEN conductor in mm^2		Minimum csa required for a main protective bonding in mm^2		
Copper conductor 1	Aluminium conductor 2	Copper conductor 3	Steel wire armouring 4	Aluminium strip armouring 5
35 or less	58 or less	10	80	16
over 35 up to 50	over 58 up to 83	16	128	26
over 50 up to 95	over 83 up to 159	25	200	41
over 95 up to 150	over 159 up to 253	35	280	57
over 150	over 253	50	400	81

Note: Local electricity distribution network conditions may require a larger main protective bonding conductor csa than listed in columns 3, 4 or 5.

Where PME conditions **apply**, Regulation 544.1.1 requires the protective bonding conductors to be selected based upon the csa of the PEN conductor of the supply using Table 54.8 of *BS 7671*. Table 2.3 is partially based on information given in Table 54.8.

Furthermore, where PME conditions apply, the requirements of Regulation 544.1.1 of the installation also applies to the earthing conductor of the installation (see Regulation 542.3.1). The csa of the earthing conductor must be not less than the minimum required for its function as an earthing conductor and for its function as a main protective bonding conductor.

The note in Table 2.3 means that the data in Table 54.8 should be used as a guide only, and that the specific requirements of the relevant electricity distributor should always be determined when considering the selection of protective bonding conductors where PME conditions apply.

Table 2.4 The requirements of *BS 7671* for minimum main protective bonding conductor sizes where PME conditions do not apply.

Csa of main line and neutral conductors, for example meter tails, in mm^2 (Note 1)	Minimum csa required for a copper earthing conductor no buried with both corrosion and mechanical protection, in mm^2 (Notes 2 & 3)	Minimum csa required for main protective bonding conductor, in mm^2 (Note 4)	
		Copper conductor	Steel wire armouring
4 or 6	6	6	24
10	10	6	40
16 or 25	16	10	64
50	25	16	100
7-0	35	25	140

1) The distributor may specify a minimum line and neutral conductor size in relation to cut-out fuse rated current.

2) The minimum earthing conductor sizes in column 2 have been selected in relation to the line conductor sizes in column 1, in accordance with Table 54.7 of *BS 7671*. Alternatively, however, the minimum earthing conductor size may be calculated in accordance with Regulation 543.1.3 (adiabatic equation).

3) Buried copper earthing conductors without mechanical protection are subject to a minimum size of 16 mm^2 where protected against corrosion or 25 mm^2 where not protected against corrosion.

4) The minimum main protective bonding conductor sizes in columns 3 and 4 have been chosen on the basis of the earthing conductor sizes in column 2, In accordance with the relevant requirements of Regulation 544.1.1.

Regulation 544.1.2 of *BS 7671* requires the main protective bonding connection to any extraneous-conductive-part, such as gas, water or other service to be made as near as practicable to the point of entry of that part into the premises. Where there is a meter, isolation point or union, the regulation further requires that the connection is made to the consumer's hard metal pipework (not to a lead pipe or a flexible pipe for example) and before any branch pipework. Where practicable, the connection should be made within 600 mm of the meter outlet union or at the point of entry of the service to the dwelling if the meter is external (Fig 2.8 refers). Where an electrical installation serves more than one building, such as a dwelling and a remote garage, these requirements should be applied at each building.

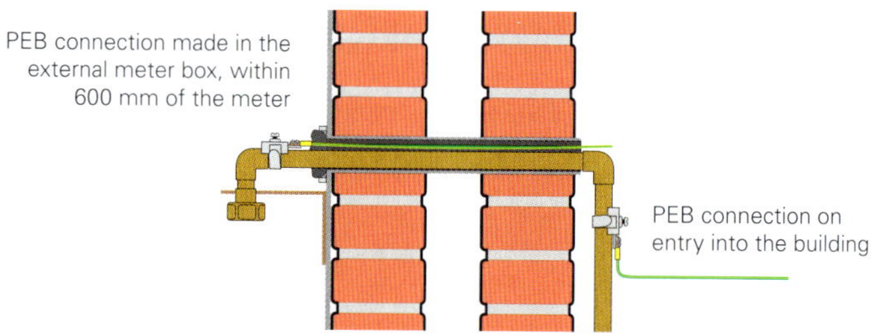

Fig 2.8 Protective bonding to the gas installation pipework

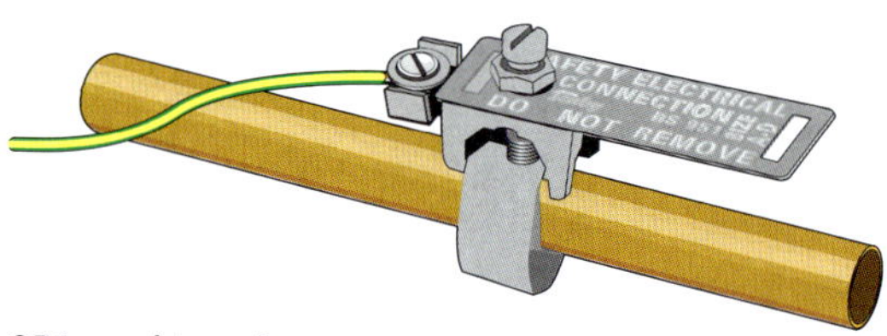

Fig 2.9 *BS 951* earthing clamp

It is generally preferable for a suitable bonding connection to be made inside the premises, but this may not always be practicable.

Where a common bonding conductor loops in and out to connect to an extraneous conductive-part, the conductor should be left unbroken at the connection. This is so that the continuity to other items connected to the bonding conductor will not be lost if the connection becomes detached (for example, as a result of work being carried out to pipework).

A permanent label bearing the words 'Safety Electrical Connection — Do Not Remove' should be installed in a visible position at or near the point of connection of every protective bonding conductor to an extraneous-conductive-part, and at the main earthing terminal where this is separate from the consumer unit or main switchgear (Regulation 514.13.1 and Fig 2.9 refer).

2.8 Supplementary bonding

Supplementary equipotential bonding ('supplementary bonding' for short) is a complementary measure for fault protection, which is required by *BS 7671* in certain situations. Supplementary bonding involves connecting together the conductive parts of electrical items and non-electrical items, such as those shown in Fig 2.10, to prevent the occurrence of a dangerous voltage between them under earth fault conditions.

BS 7671 calls for supplementary bonding to be provided:

- for certain types of special installations or locations covered by Part 7 of the Regulations
- where the conditions for automatic disconnection in case of a fault of Regulation Group 411.3.2 be fulfilled by an overcurrent device - which is not likely to be the case in a dwelling.

Those instances of increased shock risk that require supplementary bonding in a dwelling are typically locations containing swimming pools and other basins. It should be noted that supplementary bonding will be required in some cases in locations containing a bath or shower (Regulation 701.415.2 refers). See Chapter 11 of this Guide for guidance regarding special locations.

Where supplementary bonding is required, Regulation 415.2.1 requires it to connect together all simultaneously accessible exposed-conductive-parts of equipment and extraneous-conductive-parts (including, where appropriate, the main metallic reinforcement within reinforced concrete forming part of the structure). The equipotential bonding should also be connected to the protective conductors of all equipment forming a part of the installation including the earthing contacts of socket-outlets (see Fig 2.10).

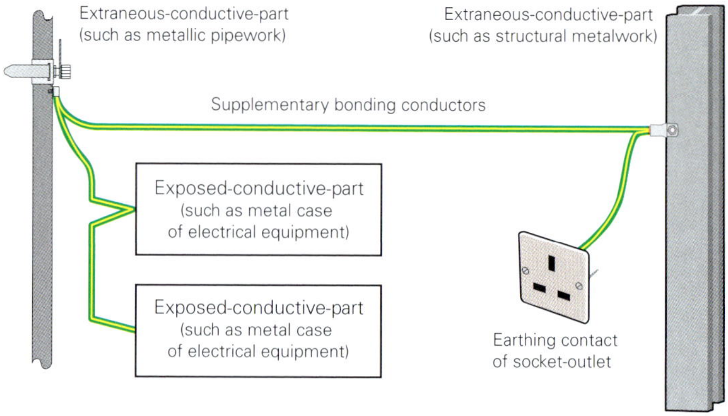

Fig 2.10 Example of items to be connected together where supplementary bonding is required

Where supplementary bonding is required because the location falls within the scope of Part 7 of *BS 7671* (Special Installations or Locations), the relevant regulations in Part 7 may require items additional to those referred to in Regulation 415.2.1 to be connected together by the supplementary bonding. For example, for a location containing a bath or shower, the terminal of the protective conductor of each circuit supplying Class I **and** Class II equipment is included amongst the items that must be connected together by supplementary bonding, as required by Regulation 701.415.2.

Single-core, non-flexible, copper cable having a green-and-yellow covering is commonly used for supplementary bonding. However, this is not the only type of conductor that may be used. For example, a protective conductor forming part of a sheathed cable (such as 'twin & earth') or flexible cable may be employed as a supplementary bonding conductor. This is subject to all the relevant requirements of *BS 7671* being met, including that the cross-sectional area of the conductor must be sufficient for such use (see Table 2.5) and for use as the circuit protective conductor where the conductor also performs that function.

Regulation 543.2.1 permits the use of other types of conductor as a supplementary bonding conductor. The types of conductor which may be used include metallic parts of wiring systems and extraneous-conductive-parts, or a combination of these, provided that all the relevant requirements of *BS 7671* are met. For example, parts of a metal pipework system such as that for cold water, hot water or heating, may provide permanent and reliable electrical continuity for supplementary bonding purposes, to the standard required by Section 543. Note, however, that the following must not be selected as a bonding conductor (Regulation 543.2.3 refers):

- gas or oil pipes,
- flexible or pliable conduit,
- support wires,
- any constructional parts that are deemed to be subject to mechanical stress in normal usage.

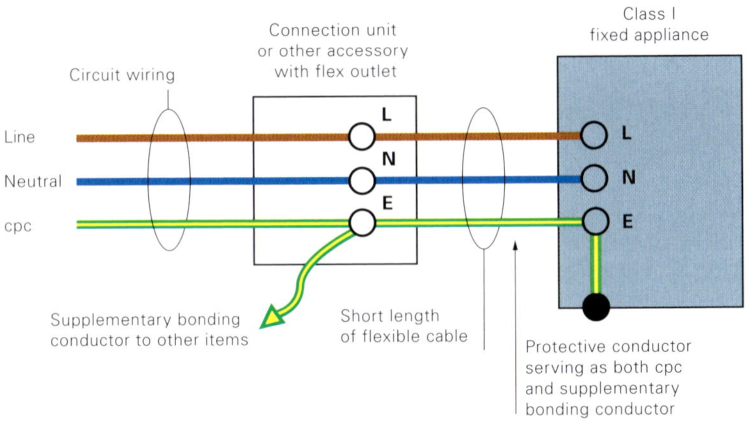

Fig 2.11 Use of cpc in a short length of flexible cable for supplementary bonding

Fig 2.11 shows the application of the requirements of Regulation 543.2.1, for a situation where supplementary bonding is required to a Class I fixed appliance supplied via a short length of flexible cable from an adjacent connection unit or other accessory incorporating a flex outlet. Regulation 544.2.5 allows the circuit protective conductor (cpc) within the flexible cable to be used to provide the supplementary bonding connection to the exposed-conductive-parts of the appliance, from the earthing terminal in the connection unit or other accessory where the supplementary bonding conductor to other items is terminated.

Table 2.5 Minimum csa of copper conductors required for supplementary bonding

Bonding and conductor connecting			
Mechanical protection? (1)	Two exposed conductive-parts	An exposed-conductive-part and an extranaeous-conductive-part	Two extranaeous-conductive-parts
Yes	csa ≥ csa of smaller cpc	csa ≥ csa of cpc	csa ≥2.5 mm^2
No	csa ≥4 mm^2	csa ≥4 mm^2	csa ≥4 mm^2

Note 1: Where sheathed* or otherwise provided with mechanical protection (for example, by encosure in conduit etc)

*The covering (insulation) of a single-core non-sheathed cable (for example, to *BS 6004* or *BS 7211*) is not a sheath.

The cross-sectional area (csa) of supplementary bonding conductors must be not less than that necessary to satisfy the 544.2 Regulation Group, as shown in Table 2.5.

Regulation 415.2.2 imposes a limit on the ohmic value of the resistance (R) of a supplementary bonding conductor that connects exposed-conductive-parts and extraneous-conductive-parts, as given in the following equation in AC systems:

$$R \leq \frac{50}{I_a}$$

Where I_a is the operating current of the protective device.

For an RCD, I_a is the rated residual operating current $I_{\Delta n}$. For an overcurrent protective device I_a is the minimum current which disconnects the circuit within 5 s.

Meeting these requirements for resistance (R) is unlikely to be a problem in practice in most small and domestic premises.

As with all conductors, supplementary bonding conductors must be adequately supported. Attention must also be paid to their ability to withstand external influences such as mechanical damage (Section 522 refers), and any anticipated factors likely to result in deterioration (Regulation 543.3.1 refers).

As with every other type of electrical connection, supplementary bonding connections and joints are required to have durable electrical continuity and adequate strength (Regulation 526.1 refers). In addition, Regulation 526.3 requires, with certain exceptions, that every connection and joint is accessible for inspection, testing and maintenance.

Supplementary bonding connections to metal pipework should normally be made using bonding clamps complying with *BS 951* (see Fig 2.9). Clamps should be selected to suit both the pipe diameter and bonding conductor size. Other suitable means of connection will be required where connecting to extraneous-conductive parts other than pipes, such as structural steelwork.

A permanent label as required by Regulation 514.13.1, bearing the words 'Safety Electrical Connection — Do Not Remove', must be permanently fixed in a visible position at or near the point of connection of every bonding conductor to an extraneous-conductive-part.

To counter a popular misconception: there is no requirement in *BS 7671* to provide supplementary bonding between a metallic kitchen sink and associated pipework. However, where a metallic waste pipe that meets the definition of an extraneous conductive-part serves the sink, a main protective bonding conductor would be required to connect it to the MET.

2.9 Sizing of circuit protective conductors

The cross sectional area (csa) of every protective conductor, other than a protective bonding conductor, must be either:

- calculated (in accordance with Regulation 543.1.3), or
- selected (in accordance with Regulation 543.1.4).

Where a circuit protective conductor forms part of a standard circuit arrangements, tables on maximum circuit length are provided in section A4 of Appendix A of this Guide.

2.10 Earthing arrangements where an installation contains equipment having high protective conductor currents

The earthing requirements for the installation of equipment having high protective conductor currents given in Regulation Group 543.7 are summarised in Table 2.6.

Table 2.6 Earthing requirements for equipment having high protective conductor currents

Ref	Application	Additional protective conductor connection requirements (alternatives)	Regulation
1	Connection of equipment having protective conductor current exceeding 3.5 mA but not exceeding 10 mA.	Permanent connection to installation wiring.	543.7.1.201
		Connection by means of *BS EN 60309-2* a plug and socket	
2	Connection of equipment having protective conductor current exceeding 10 mA.	Permanent connection to installation wiring. Protective conductor selected in accordance with Regulation 543.7.203 (see Ref 3)	543.7.1.202
		Flexible cable with *BS EN 60309-2* plug and socket. Csa of protective conductor of flexible cable not less than: • 2.5 mm^2 for plugs rated at 16 A, or • 4 mm^2 for plugs rated above 16 A, or • csa of line conductor.	
		Protective conductor complying with Section 543 with earth monitoring system to *BS 4444*.	
3	Final circuit or distribution circuit supplying one or more items of equipment, the total protective conductor being likely to exceed 10 mA (provision of high integrity protective connection)	A single copper protective conductor of csa not less than 10 mm^2, complying with Regulations 543.2 (types of protective conductor) and 543.3 (preservation of continuity)	543.7.1.203 and 543.7.1.204
		Two individual protective conductors, which may be of different types, such as a metallic conduit and an additional conductor enclosed within. Each must comply with Section 543 (Protective conductors).	
		The ends of the two protective conductors must be terminated independently of each other at all connection points throughout the circuit, such as distribution board, junction boxes and socket-outlets. (Regulation 543.7.1.204 refers).	
		If both protective conductors are incorporated in a multicore cable, total csa of all the conductors, including live conductors, to be not less than 10 mm^2. One protective conductor may be formed by metallic sheath, armour or braid of the cable, provided this complies with Regulation 543.2.5	
		Protective conductor complying with Section 543 with earth monitoring system to *BS 4444*.	
		Connection of the equipment to the supply by double wound transformer or equivalent, such as a motor-alternator set. See Regulation 543.7.1.203(v) for detailed requirements.	

4	Final circuit intended to supply two or more items where total protective conductor current in normal service is likely to exceed 10 mA (provision of high integrity protective connection)	Ring final circuit with a ring protective conductor. Spurs, if provided, require high integrity protective conductor connection complying with requirements of Regulation 543.7.1 (see Ref 3)	543.7.1.201
		The radial final circuit with a single protective conductor, provided that also: • the protective conductor is connected as a ring, or • a separate protective conductoris provided at the final socket-outlet by connection to the metal conduit or ducting, or • in the case where a number of similar radial circuits supply socket-outlets in adjacent areas, a second protective conductor may be provided at the final socket-outlet on one circuit by connection to the protective conductor of the adjacent circuit. The circuits must be fed from the same distribution board, have identical means of short circuit and overcurrent protection and circuit protective conductors of the same csa.	
		Other circuits complying with the requirements of Regulations 543.7.1.201 to 543.7.1.204 (see Refs 1,2 and 3)	

CHAPTER 3
PROTECTIVE DEVICES

3.1 Introduction to protective devices

Unless Regulation 434.3 applies, a protective device(s) shall be installed in every circuit and shall be of an appropriate type in accordance with Section 432.

The following protective devices are commonly available.

- Fuses to *BS 88-2*
- Fuses to *BS 88-3*
- Fuses to *BS 1362*
- Circuit-breakers to *BS EN 60898* (as amended)
- Residual current operated circuit-breakers without integral overcurrent protection (RCCB) to *BS EN 61008*
- Residual current operated circuit-breakers with integral overcurrent protection (RCBO) to *BS EN 61009*
- Type F and type B residual current operated circuit-breakers for household and similar uses to *BS EN 62423*

The following protective devices are no longer in production but are likely to be encountered in some existing domestic premises.

- Rewireable (semi-enclosed) fuses to *BS 3036*
- Cartridge fuses to *BS 1361*
- Circuit-breakers to *BS 3871*.

All listed protective devices will be discussed in this Chapter. *BS 7671* now contains requirements relating to overvoltage control. These are also considered in relation to domestic and similar premises.

3.2 Cartridge fuses

3.2.1 *BS 88*

Cartridge fuses to *BS 88* are widely used as circuit protective devices in commerce and industry, less so in domestic premises.

There are a number of types of *BS 88* fuse (see Fig 3.1). The general requirements for *BS 88* fuses are now specified in *BS 88-1* (also known as *BS EN 60269-1*), and the supplementary requirements for *BS 88* fuses intended for specific conditions of use or applications are covered in the subsequent Parts of the *BS 88* series.

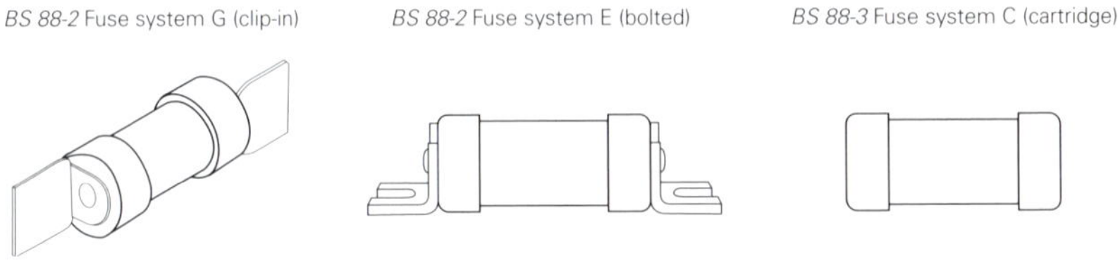

Fig 3.1 Cartridge fuses to *BS 1361*.

BS 7671 contains information relating to the maximum permitted earth fault loop impedance and time / current characteristics for *BS 88-2* fuse systems E and G, which are suitable for use by authorised persons, and to *BS 88-3* fuse system C, which may be used or replaced by unskilled persons, such as householders.

Fuse system C type fuses are sub-divided into two types - type I and type II. Details of these are given in Table 3.1.

Table 3.1 Information relating to *BS 88-3* system C fuses.

Type	Rated current	Rated voltage	Rated breaking capacity
I	≤ 45 A	230 V	16 kA
II	≤ 100 A	400 V	31.5 kA

3.2.2 *BS 1361*

Up until the publication of Amendment No 1 (2011) to the 17th Edition of the Wiring Regulations, *BS 7671* contained data relating to cartridge fuses conforming to *BS 1361*. Broadly speaking this type of fuse has been replaced by those to *BS 88-3* fuse system C, described above.

There are two basic types of *BS 1361* fuse - Type I and Type II. Type I fuses are intended for use in distribution boards where the fuses might be replaced by the consumer. Type II fuses are primarily intended for use as distributor's cut-out fuses, in domestic and similar premises.

BS 1361 fuses are colour-coded as follows:

Table 3.2 Information relating to *BS 1361* fuses.

5 A – white	15 A – blue
20 A – yellow	30 A – red
45 A – green	

Table 3.3 Information relating to *BS 1361* fuses.

Type	Rated current	Rated voltage	Rated breaking capacity
I	5 to 45 A	240 V	16 kA
IIa	60 A and 80 A	415 V	31.5 kA
IIb	80 A and 100 A		

3.2.3 *BS 1362*

Cartridge fuses conforming to *BS 1362* have a rated voltage of 240 V and are suitable for use in low voltage single-phase AC circuits of nominal frequency 50 or 60 Hz. These are general-purpose fuses intended for domestic and similar purposes, primarily in plugs complying with *BS 1363*, although not exclusively so. *BS 1362* cartridge fuses are, for example, commonly used in fused connection units.

The rated current (I_n) allocated by the manufacturer may be any value up to and including 13 A, although the preferred values in *BS 1362* are 3 A and 13 A.

Where a fuse is used to provide protection against fault current only, for example to protect the flexible cable to an appliance with integral overload protection, the fault clearance time under both short-circuit and earth fault conditions has to meet the requirements of Regulation 434.5.2.

For overload current protection, the rated current of a fuse is to be selected to meet the requirements of Section 433 of *BS 7671*.

Where overload current protection is afforded by a *BS 1362* cartridge fuse, there tends to be a need for the circuit conductors to be of a larger cross-sectional area than required for most other types of overcurrent protective device used. This is because of the need to comply with condition (iii) of Regulation 433.1.1, which requires that the current causing effective operation of the protective device (I_2), does not exceed 1.45 times the lowest of the current-carrying capacities I_z of any of the conductors of the circuit.

BS 1362 cartridge fuses should be indelibly marked with the following information.

- Manufacturer's name or identifying mark
- British Standard Number
- Rated current in amperes.

The fuse-link barrel should also be colour coded according to rated current as follows: brown for 13 A fuses, red for 3 A fuses and black for all other current ratings (such as 2 A, 5 A and 10 A).

3.3 Semi-enclosed fuses to *BS 3036*

Requirements applicable to rewireable (semi-enclosed) fuses are specified in *BS 3036: 1958 (2007) Semi-enclosed electric fuses.*

BS 3036 defines the fusing factor of a fuse covered by that standard as: 'The ratio, greater than unity, of the minimum fusing current to the current rating of a fuse', and is not to exceed 2.0.

BS 7671 recognises that the minimum fusing current, I_2 for a *BS 3036* fuse can be as high as 2.0 I_n and advises that, where such a fuse is used for overload current protection, compliance with Regulation 433.1.1 is afforded where the rated current, I_n of the fuse is selected such that it does not exceed 0.725 times the current-carrying capacity I_z of the lowest rated conductor in the circuit protected (Regulation 433.1.202 refers).

Where a semi-enclosed fuse is selected, it shall be fitted with an element in accordance with the manufacturer's instructions, if any. In the absence of such instructions, it shall be fitted with a single element of tinned copper wire of the appropriate diameter specified in Table 533.1 in *BS 7671*

3.4 Circuit-breakers

3.4.1 *BS EN 60898*

Circuit-breakers to *BS EN 60898-1: 2003 +A13: 2012 – Circuit-breakers for overcurrent protection for household and similar installations.* Circuit-breakers for AC operation (see Fig 3.2), are widely used as final circuit protective devices in distribution boards, consumer units and control panels.

This Guide gives information on the main characteristics of such circuit-breakers as far as fault protection, protection against overcurrent, and isolation and switching are concerned.

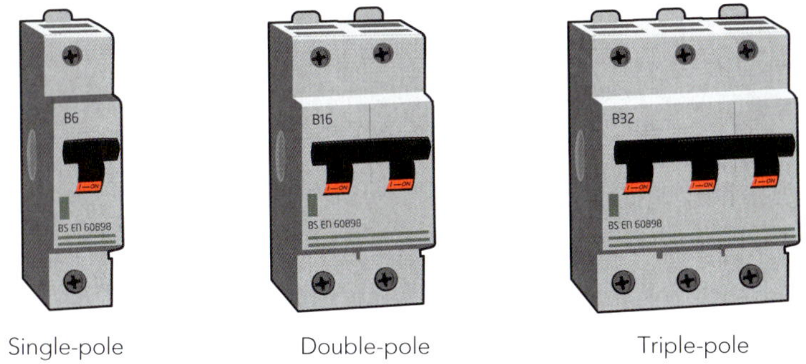

Fig 3.2 Typical circuit-breakers to *BS EN 60898*.

The preferred values of rated current given in *BS EN 60898-1* are 6, 8, 10,13, 16, 20, 25, 32, 40, 50, 63, 80, 100 and 125 A. Rated current I_n, is the current that a circuit-breaker is designed to carry indefinitely at a specified reference ambient air temperature, normally of 30 °C.

Circuit-breakers conforming to *BS EN 60898-1* are classified into three distinct 'Types'; Type B, Type C and Type D, according to their instantaneous tripping current.

Instantaneous tripping current is the minimum value of current causing a circuit-breaker to operate automatically without intentional time delay - that is, within a time not exceeding 0.1 s (100 ms). The limiting value of earth fault loop impedance (Z_s) for a circuit-breaker for fault protection, as given in Chapter 41 of *BS 7671*, is based on the instantaneous tripping current of the device.

BS EN 60898-1 permits instantaneous tripping to begin to occur at a value of fault current, expressed as a multiple of the rated current (I_n), within a range of values for each circuit-breaker type. The standard ranges are shown in Table 3.4, although manufacturers may design circuit-breakers to operate instantaneously within smaller ranges of current, falling within the standard ranges.

Table 3.4 Instantaneous tripping

Circuit-breaker type	Standard range	
B	above 3 I_n	up to and including 5 I_n
C	above 5 I_n	up to and including 10 I_n
D	above 10 I_n	up to and including 20 I_n

The standard ranges of instantaneous tripping current are also shown in graphical form in Fig 3.4, which is a representation of the time / current characteristics of Type B, C and D circuit-breakers.

Generally, in order to meet the requirements of *BS 7671* for automatic disconnection to occur within a prescribed time limit (such as, say, 0.2 s or 1 s, as appropriate) under earth fault conditions, the prospective earth fault current must be not less than the upper value of the range of instantaneous tripping currents for the circuit-breaker concerned.

For example, on the basis of the standard ranges of current referred to in Table 3.4 and Fig 3.3, the prospective earth fault current for a 32 A type B circuit-breaker is required to be not less than 160 A (that is, $5I_n$). Correspondingly, prospective earth fault currents of not less than 320 A and 640 A, respectively, are required for instantaneous tripping of 32 A type C and D circuit-breakers.

The higher the prospective fault current that is required to cause instantaneous tripping, the lower the limiting earth fault loop impedance (Z_s) needs to be.

Fig 3.3 Graphical representation of time / current characteristics of type B, C and D circuit-breakers (including the ranges of current within which instantaneous tripping begins to occur)

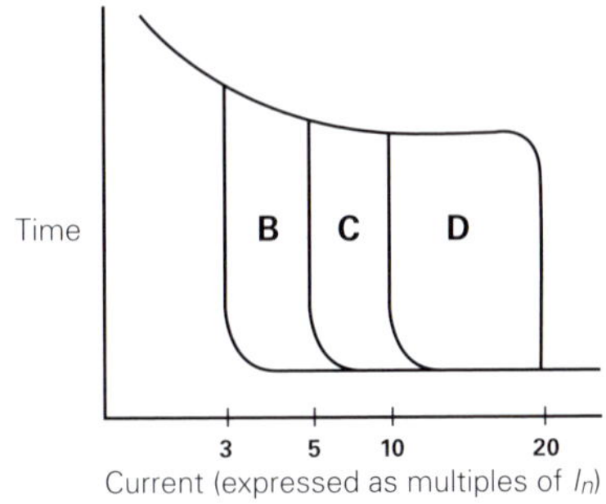

This can be seen from the limiting values of earth fault loop impedance given in Chapter 41 of *BS 7671*, for types B, C and D circuit-breakers.

Due to the particular time / current characteristics of a circuit-breaker, values of earth fault loop impedance even marginally exceeding the limiting values in the above tables of *BS 7671* may result in greatly increased disconnection times which would not be acceptable for fault protection.

As it cannot be predicted where in the range a circuit-breaker will trip, the limiting values of earth fault loop impedance for *BS EN 60898* circuit-breakers, in Chapter 41 are based on the upper range values of current indicated by Table 3.4 earlier. The limiting values are derived by the method given in the introduction to Appendix 3 of *BS 7671* (time / current characteristics of overcurrent protective devices).

Manufacturers of *BS EN 60898* circuit-breakers sometimes publish, in relation to their own circuit-breakers only, limiting values of earth fault loop impedance higher than the corresponding limiting values given in Chapter 41. This is because, as indicated earlier in this Chapter, manufacturers are free to design circuit-breakers to operate instantaneously within ranges of current smaller than, but falling within, the standard ranges given in Table 3.4.

Furthermore, some manufacturers of *BS EN 60898* circuit-breakers publish limiting values of earth fault loop impedance to facilitate tripping in a time of up to 5 s (that is, in addition to limiting values of earth fault loop impedance based on instantaneous tripping current). This is because the characteristics of the circuit-breakers concerned can be such as to allow tripping other than instantaneously without the risk of a greatly increased disconnection time.

The rated short-circuit capacity (I_{cn}) of a circuit-breaker conforming to *BS EN 60898* is the value of the ultimate short-circuit capacity assigned by the manufacturer.

A circuit-breaker also has a corresponding service short-circuit capacity I_{CS}, which is either equal to or lower than the rated short-circuit capacity (I_{CS}), depending on the I_{cn} of the device.

Where a circuit-breaker has cleared a short-circuit fault or earth fault at a prospective current in excess of its rated value of I_{CS}, the device is likely to be unfit for continued service and should be replaced. Of the two short-circuit capacities (I_{CS} and I_{cn}), only the rated short-circuit capacity (I_{CS}) is required to be marked on a circuit-breaker. This is shown within a rectangle, as illustrated in Fig 3.4. The example indicates a rated short-circuit capacity of 6000 A.

6000

Fig 3.4 Example of marking that denotes the rated (ultimate) short-circuit capacity (I_{cn})

Within the scope of this Guide, the preferred values of rated short-circuit capacity (I_{CS}) given in *BS EN 60898-1* are 1500 A, 3000 A, 4500 A, 6000 A and 10000 A.

3.4.2 *BS 3871*

Although *BS 3871: Part 1: 1965 - Specification for miniature and moulded case circuit-breakers. BS 3871 MCBs* may still be encountered when carrying out work in existing installations.

The short-circuit capacity of a *BS 3871* circuit-breaker is indicated by a capital 'M' followed by the rated duty in kilo-amps (kA). For example, if the short-circuit capacity of a particular circuit-breaker was 3000 A this would be identified as M3 on the circuit-breaker.

BS 3871 circuit-breakers were available in a range of types according to their instantaneous tripping current, as given in Table 2 of that standard, and shown in Table 3.5.

Table 3.5 Instantaneous tripping current by type for *BS 3871* MCBs

Circuit-breaker type	Instantanaeous tripping current	
1	above 2.7 I_n	up to and including 4 I_n
2	above 4 I_n	up to and including 7 I_n
3	above 7 I_n	up to and including 10 I_n
4	above 10 I_n	up to and including 50 I_n

3.5 Residual current devices

A Residual current device (RCD) monitors the algebraic sum of the currents in the line and neutral conductors (the residual current) at the point where it is installed in a circuit, and mechanically interrupts the circuit when this sum exceeds a preset value (the rated residual operating current, $I_{\Delta n}$).

The following definition is given in Part 2 of *BS 7671*.

> **Residual current device (RCD).** A mechanical switching device or association of devices intended to cause the opening of the contacts when the residual current attains a given value under specified conditions.

The term RCD therefore includes:

- RCCB (residual current operated circuit-breaker without integral overcurrent protection),
- RCBO (residual current operated circuit-breaker with integral overcurrent protection),
- SRCD (socket-outlet incorporating an RCD),
- Portable RCD - an RCD formed of a suitable association of devices including current transformers, a residual current sensor, a test button and a circuit-breaker.

3.5.1 RCCBs to *BS EN 61008* and *BS 4293*

A residual current operated circuit-breaker without integral overcurrent protection (RCCB) is not designed to perform the function of protection against overcurrent (overload current or fault current).

A circuit, controlled by an RCCB must therefore be provided with a means for protection against overcurrent, such as suitable circuit-breakers or fuses.

RCCBs should comply with *BS EN 61008-1: 2012 +A12: 2017- Residual current operated circuit-breakers without integral overcurrent protection for household and similar uses (RCCBs) – Part 1: General rules (IEC 61008-1: 2010, modified).*

Although *BS 4293: 1983 Specification for Residual current-operated circuit-breakers* was withdrawn in December 2009, *BS 4293* RCDs might still be encountered when carrying out work in existing installations. For this reason, information relating to the requirements for their testing is included in Chapter 15 of this Guide.

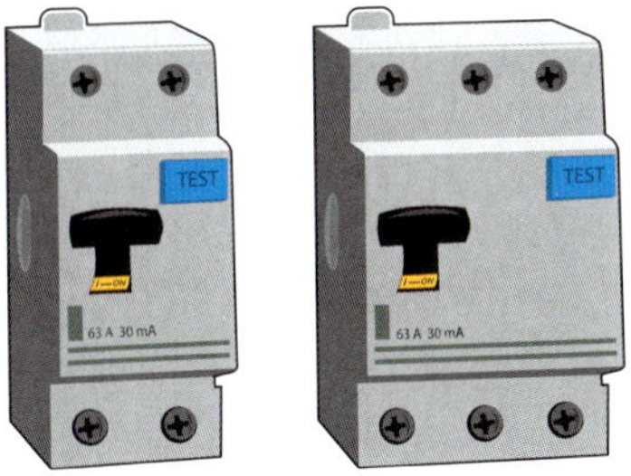

Fig 3.5 RCCBs for DIN rail mounting for use in an enclosure such as consumer unit.

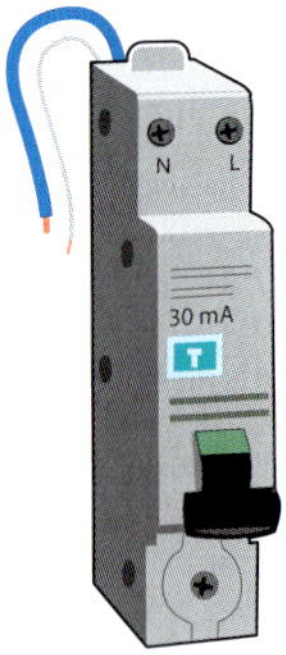

Fig 3.6 Typical RCBO

3.5.2 RCBOs to *BS EN 61009*

A residual current operated circuit-breaker with integral overcurrent protection (RCBO) is designed to perform the function of protection against overcurrent, as well as that of an RCD.

RCBOs are often used as circuit protective devices in consumer units and distribution boards (see Fig 3.6).

RCBOs should comply with *BS EN 61009-1: 2012 +A2: 2014 Residual current operated circuit-breakers with integral overcurrent protection for household and similar uses (RCBOs) — Part 1: General rules*

3.5.3 Classification of RCDs (RCD type)

When used to describe the operating characteristics of an RCD, the term 'type' relates to the frequency and shape of current waveform for which tripping is assured.

In alternating current circuits, particularly those supplying electronic equipment, even though the voltage is a sine wave, the current flowing in the circuit can be non-sinusoidal and may have a DC component. In domestic installations the increased use of electronic devices, in particular the use of electric vehicle charging systems and inverters in micro-generators systems, means that many circuits now draw a non-sinusoidal current.

When RCDs of the type designed to operate with a sinusoidal current are used in circuits supplying electronic equipment that draw a non-sinusoidal current then the RCD may not function satisfactorily.

To provide installers of RCDs with guidance as to the ability of an RCD to function satisfactorily when carrying current with specified waveforms, product standard committees have developed a number of standard performance criteria known as type tests. These performance criteria are used to confirm the performance of the RCD is of a certain 'type'; the 'type' being indicated by using the classifications AC, A, B or F.

The British Standard applicable to Type B and F RCDs is *BS EN 62423: 2012*. For each of the standard RCD 'types', in addition to providing description of the current waveforms for which tripping is assured Table 3.6 shows the symbols used to provide the user with a visual indication of the particular current waveforms.

Type AC RCDs, the type of RCD used in most electrical installations may not operate satisfactorily when used in circuits supplying electronic equipment. For example, Sections 710

(Medical locations), Section 712 (Solar photovoltaic power supply systems) and Section 722 (Electric vehicle charging installations) of *BS 7671* all include requirements for the use of RCDs other than type AC.

Table 3.6 Suitability of various types of RCD

Type	Operation is assured for
AC	residual sinusoidal alternating currents, whether suddenly applied or slowly rising.
A	as for type AC and, in addition, residual pulsating direct currents and residual pulsating direct currents superimposed on a smooth direct current of 6 mA.
B	as for type A, and in addition: residual sinusoidal alternating currents up to 1000 Hz, residual alternating currents superimposed on a smooth direct current of 0.4 times the rated residual operating current, residual pulsating direct currents superimposed on a smooth direct current of 0.4 times the rated residual operating current and residual direct currents which may result from rectifying circuits.
F	as for type A, and for residual currents from mixed frequencies up to 1000 Hz

3.5.4 RCD protected socket-outlets

SRCDs should comply with *BS 7288: 2016 Specification for residual current devices with or without protection for socket-outlets for household and similar uses.*

The use of an SRCD does not preclude the need to provide RCD protection to the final circuit supplying the SRCD where this is required by *BS 7671* (such as for the purposes of Regulation 522.6.202 or 522.6.203), relating to cables concealed in a wall or partition, or for one of the other purposes mentioned in Table 3.7 of Section 3.5.5 of this Chapter.

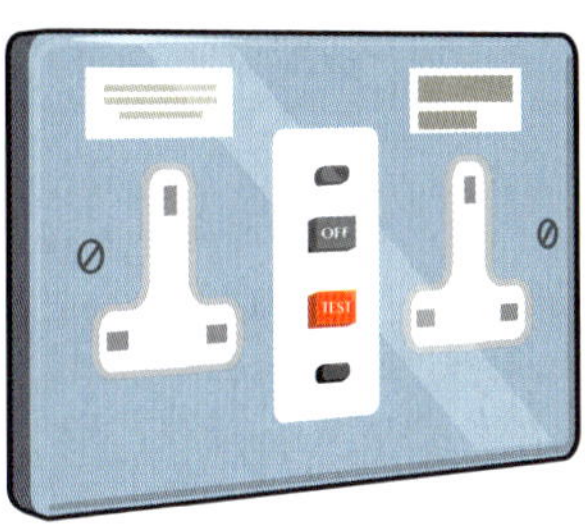

Fig 3.7 Example of an SRCD (socket-outlet with integral RCD)

3.5.5 Typical applications for RCDs in domestic premises

Table 3.7 summarises the typical applications and rated residual currents for RCDs as required by *BS 7671* that are applicable to installations in domestic and similar premises.

Table 3.7 Applications for RCDs in domestic premises.

Application	Requirement	Relevant *BS 7671* regulation(s)
Additional protection (against electric shock) – general, ADS	$I_{\Delta n}$ = 30 mA	411.3.3; 411.3.4, 415.1.1; 415.1.2
Fault protection shock – General, ADS, in TN system	As required in order to achieve required disconnection time $I_{\Delta n}$ = 30; 100; 300; 500 mA	411.4.204
Fault protection – general, ADS, in TT system Note: RCD protection is preferred for fault protection in TT systems (411.5.2 refers)	As required in order to achieve required disconnection time, and to ensure that $R_A \times I_{\Delta n} \leq 50$ V	411.5.3
Additional protection – cables without mechanical protection concealed in walls and partitions	$I_{\Delta n}$ = 30 mA	522.6.202; 522.6.203

Additional protection – all low-voltage circuits either serving the location containing a bath or a shower or passing through zones 1 or 2 but not serving the location	$I_{\Delta n}$ = 30 mA	701.411.3.3
Additional protection – all circuits of a sauna (except sauna heater unless specifically stated as being required by the manufacturer)	$I_{\Delta n}$ = 30 mA	703.411.3.3
Additional protection of circuit supplying equipment of floor and ceiling heating systems of Class II construction or equivalent insulation	$I_{\Delta n}$ = 30 mA	753.415.1
Solar photovoltaic installations	Where the inverter does not provide simple separation and RCD protection is required to provide either fault protection or additional protection a type B RCD is required. (Where the inverter is not able to feed fault current into the electrical installation a type B device is not required)	712.411.3.2.1.2
Electric vehicle charging installations	Every charging point to be individually protected by a 30 mA RCD that shall interrupt all live conductors. The RCD shall be a type A device unless the current has a DC component exceeding 6 mA when the RCD should be type B device.	722.531.2.101

3.6 Arc fault detection devices (AFDDs)

Arc fault detection devices conforming to *BS EN 62606*, are designed specifically to provide protection against arc faults in AC circuits (Regulation 421.1.7 refers). They are not a replacement for overcurrent protective devices or RCDs.

An AFDD is a supplementary device that is designed to differentiate between the types of arcing that occurs normally within an installation such as that produced during switching, and those dangerous electric arcs generated under fault conditions.

Typically, AFDDs are available in three formats:

a) Consisting of an arc fault detection unit (AFD unit) designed for being mechanically assembled on site in accordance with the manufacturer's instructions to one or more declared protective device(s) such as a circuit-breaker, RCBO or RCCB as shown in Fig 3.8 (a), The manufacturer shall declare which types of AFD units and protective device(s) are suitable to be assembled together or

b) As one single device, comprising an AFD unit integrated with a protective device such as a circuit-breaker, RCBO or RCCB as shown in Fig 3.8 (b); or

c) As a 'standalone' device comprising an AFD unit and opening means; as shown in Fig 3.8(c). Standalone AFDDs must be connected in series with an appropriate overcurrent protective device as declared by the AFDD manufacturer such as a circuit breaker, RCBO or fuse.

Each AFDD is designed to protect one circuit only. They are not designed to be placed immediately after the main switch for full installation arc fault protection (Regulation 532.6 refers).

Fig 3.8 (a) AFD Unit designed for being mechanically assembled on site

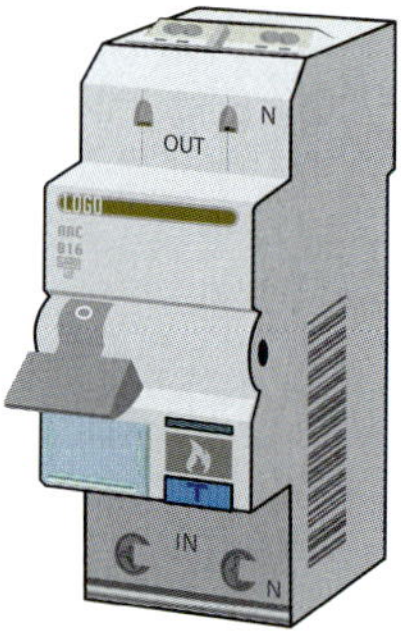

Fig 3.8 (b) AFD unit integrated with a protective device

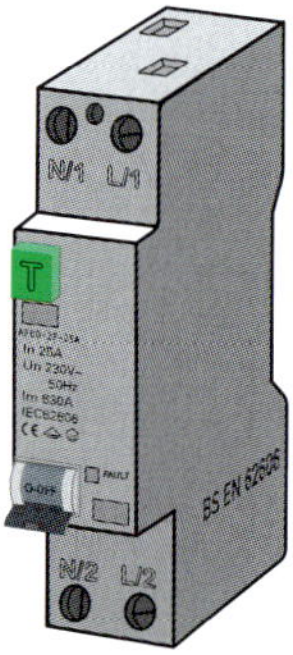

Fig 3.8 (c) Standalone device comprising an AFD unit and opening means

3.6.1 Testing AFDDs

AFDDs are provided with:

- a manual test button. When tested manually, the AFDD should trip, and / or
- an automatic test function that checks the arc detection circuit. The automatic test function consists of a test at switch-on and at intervals not exceeding at least once a day. During this automatic testing, the AFDD does not trip unless a malfunction is detected. In case a malfunction is detected the AFDD will trip and indicate a malfunction.

AFDDs integrated with an RCCB or RCBO will at least include a test button for the RCD element of the device.

3.7 Overvoltage protection and surge protective devices (SPDs)

The function of a SPD is to limit transient overvoltages of atmospheric origin and divert current to Earth. This will limit the amplitude of the overvoltage to a value that is not hazardous for the electrical installation and any connected sensitive loads.

The note to Regulation 534.4.4.4 indicates that there are two distinct causes of overvoltage:

- Transient overvoltage ('spike') – is very short pulse of energy (measured in joules) on a supply line. A transient overvoltage of atmospheric origin, such as lightning strikes, can contain very high voltages but usually last only a few milliseconds.
- Voltage surge – is a temporary increase in overvoltage due to switching on a supply line. Such surges typically have less voltage than a transient overvoltage but can last much longer.

The type of overvoltage and where it occurs has to be considered for the correct selection of SPDs with regard to their installed location. However, there is an exception not to provide protection for single dwelling units in certain situations.

Section 443 does not consider direct lightning strikes on the low voltage lines of the public supply network or to the electrical installation, see Regulation 443.1.1 and *BS EN 62305-1: 2011 Protection against lightning – general principles*, for more detail.

There are three types of SPD used in the protection of AC power circuits:

- Type 1 – this type is often referred to as an equipotential bonding SPD and is fitted at the origin of the electrical installation to specifically prevent dangerous sparking arising from lightning strikes which could lead to fire or electric shock hazards (Regulation 534.4.4.4.2 refers).

- Type 2 – this type is intended for installation on the load side of the main service equipment overcurrent protective device. Type 2 SPDs may or may not require an overcurrent protective device(s) (OCPD). Type 2 SPDs are required to protect sensitive and critical equipment (for example, hospital equipment and fire / security alarm systems) downstream of the origin of the electrical installation (Regulation 534.4.4.4.1 refers).
- Type 3 – these are called 'point of utilisation SPDs' and are installed after a Type 2 for localised over voltage protection. A surge protected extension lead would be an example of a Type 3 device.

For domestic properties supplied by a single-phase TT or TN-S earthing system, a two-pole device (single phase) would be sufficient. For a TN-C-S system, a single-pole SPD may be sufficient as it is only the live conductor that enters the property which is liable to transmit any external influence.

Appendix 16 provides further guidance on the connection of SPDs within an installation.

3.7.1 Risk assessment method

BS 7671: 2018 no longer uses the AQ criteria method, instead protection against transient overvoltage has to be provided where the consequence caused by overvoltage affects human life, public services and cultural heritage, and commercial or industrial activity.

This method recognises four consequential levels of protection, in which the consequences considered relate to:

1. human life (continuity of operation of medical equipment etc)
2. public services (effects on transport infrastructure, public services and the like)
3. commerce and industry (loss / disruption of business activities)
4. large number of individuals

Regulation 443.5 gives a formula that can be applied to designs for installations to determine whether overvoltage protection is required for a particular property.

Annex A443 provides examples on the calculated risk level (CRL). If the risk assessment and hence the CRL is not performed, the electrical installation shall be provided with protection against transient overvoltage.

For single dwelling units where the total value of the installation and equipment therein does not justify such protection, SPDs need not be provided (Regulation 443.4 refers).

CHAPTER 4
PROTECTION AGAINST ELECTRIC SHOCK

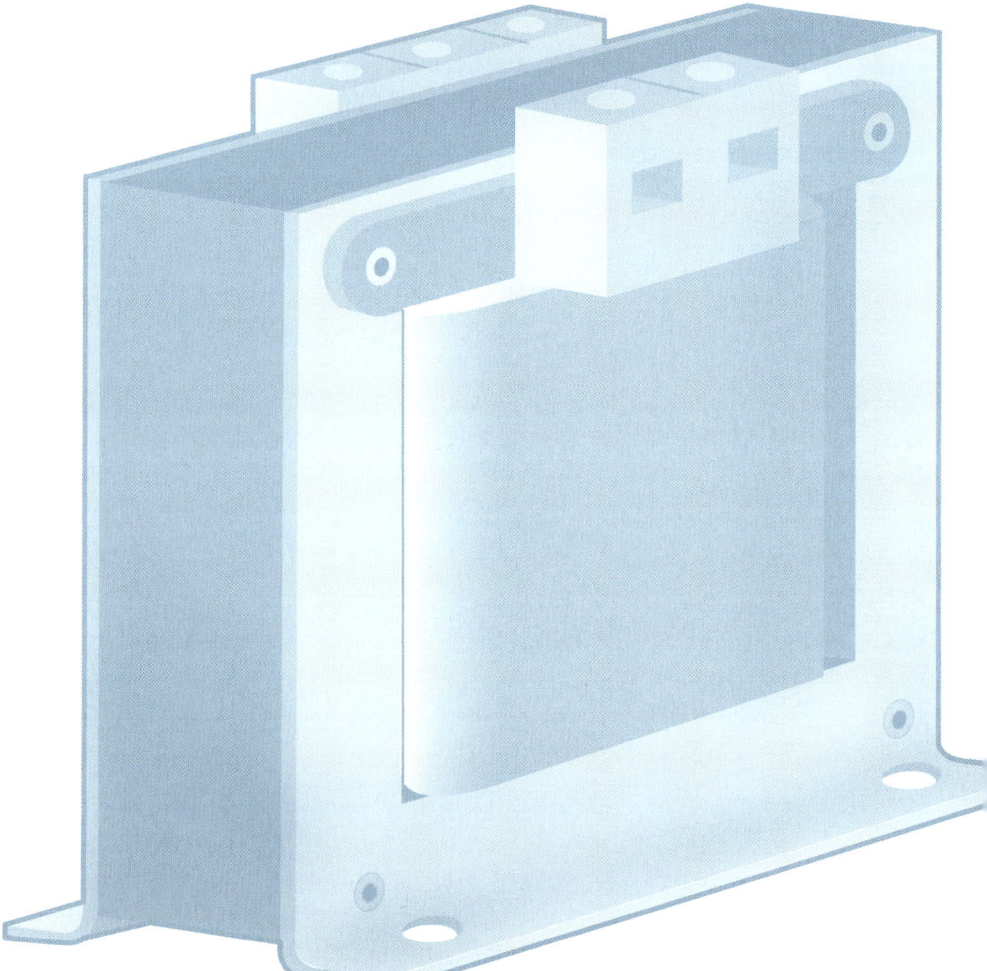

4.1 General

The fundamental rule of protection against electric shock is that:

- hazardous-live parts, such as energised conductors, must not be accessible and
- conductive parts which are accessible, such as metal enclosures of equipment or metal pipes, must not be hazardous-live.

The two conditions, above, must be achieved both in normal (no fault) conditions and under single fault conditions (such as a fault between a live conductor and a metal casing). Protection under normal conditions is achieved by basic protection.

Basic protection is defined as:

> Protection against electric shock under fault-free conditions and is provided to protect persons or livestock coming into direct contact with live parts.

A live part is defined as:

> A conductor or conductive part intended to be energised in normal use, including a neutral conductor but, by convention, not a PEN conductor.

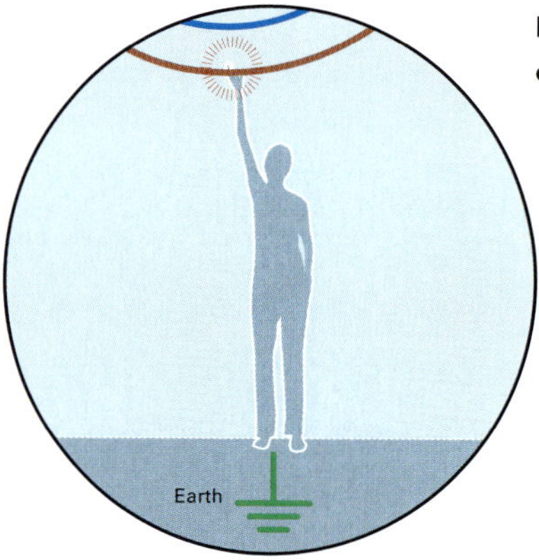

Fig 4.1 A person coming into contact with a live conductor whilst in contact with Earth

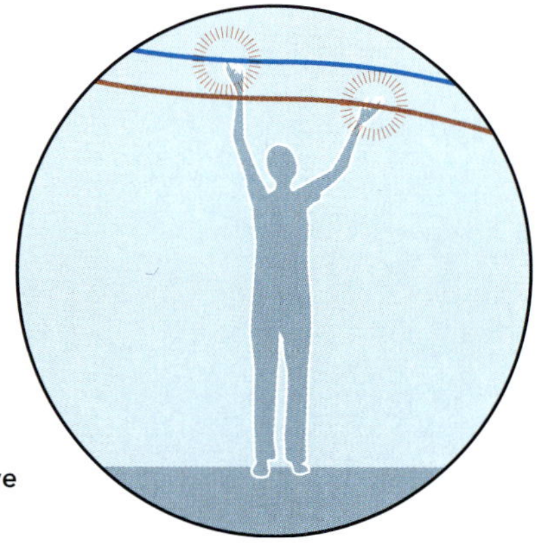

Fig 4.2 A person coming into contact with live conductors at different potentials

Fault protection provides protection for persons or livestock coming into contact with exposed-conductive-parts which have become live under single fault conditions. An exposed-conductive-part is defined as 'a conductive part of equipment which can be touched and which is not a live part but which may become live under fault conditions.' Fig 4.3 illustrates how a person could receive an electric shock under single fault conditions.

The person in Fig 4.3 is in contact with the metal enclosure of an item of Class I electrical equipment which has become live under earth fault conditions. The potential of the metal enclosure is higher than that of the main earthing terminal of the installation (and that of Earth), because of a potential difference created by the passage of earth fault current through the impedance of the circuit protective conductor.

Fig 4.3 Single fault conditions

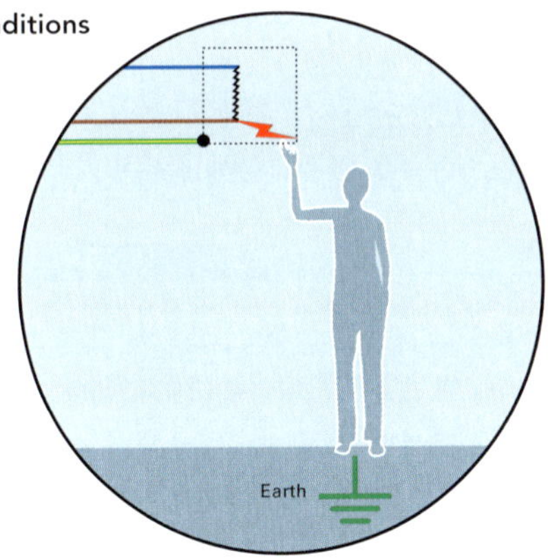

Regulation 410.3.3 recognises ADS, double or reinforced insulation, electrical separation to supply a single item of current-using equipment and extra-low voltage (SELV and PELV) as protective measures. ADS is the measure that is employed in almost every electrical installation.

A protective measure, such as Automatic Disconnection of Supply (ADS) or double or reinforced insulation must consist of a provision for basic protection and a provision for fault protection.

In the case of ADS, basic protection is provided by insulation and / or barriers and enclosures while fault protection is provided by protective earthing, protective bonding and automatic disconnection in the case of a fault (see Regulation 411.1). As can be seen, the basic and fault protective provisions are independent.

In the case of reinforced insulation, basic protection and fault protection are both provided by the reinforced insulation between live and accessible parts (Regulation 412.1.1 (ii) refers).

4.2 Provisions for basic protection

Basic protection, or protection against contact with live parts, provides protection under normal conditions and is provided by insulation and / or barriers or enclosures.

Where insulation provides basic protection it must completely cover the live part.

Such insulation must only be able to be removed by destruction, for example, by stripping the insulation from an insulated conductor prior to its termination (Regulation 416.1 refers).

Where barriers or enclosures provide basic protection they must:

- provide a degree of protection of at least IPXXB or IP2X, with permitted exceptions for access and replacement of parts (See Regulations 416.2.1 and 416.2.4),
- provide a degree of protection of IPXXD or IP4X for any readily accessible horizontal top surface (Regulation 416.2.2 refers),
- be firmly secured in place and have sufficient stability and durability taking into account the external conditions (Regulation 416.2.3 refers).

4.3 Protective measure: Automatic disconnection of supply (ADS)

The protective measure of ADS is employed in almost every electrical installation and consists of basic protection, fault protection and, for some circuits and locations, additional protection.

- Basic protection is provided by basic insulation of live parts and / or by barriers or enclosures (See 4.2).
- Fault protection is provided by protective earthing, protective equipotential bonding and automatic disconnection in the event of a fault (Regulation 411.1 refers) and is discussed further in this Chapter.
- Additional protection by one or more RCDs having a rated residual operating current ($I_{\Delta n}$) not exceeding 30 mA must be provided for specified circuits (see Regulation Group 522.6, Regulations 411.3.3, 411.3.4, 701.411.3.3 and 703.411.3.3 for example).

4.3.1 Protective earthing

Protective earthing requires all exposed-conductive-parts to be connected to a protective conductor which in turn is connected to the main earthing terminal and hence, via the Earthing Conductor to the means of earthing. In addition, a protective conductor must be run to and terminated at each point in the wiring and at each accessory, except a lampholder having no exposed-conductive-parts and suspended from such a point. (Regulation 411.3.1.1 refers).

4.3.2 Protective equipotential bonding

In each installation, protective equipotential bonding conductors (formerly referred to as main bonding conductors) complying with Chapter 54, are required to connect extraneous-conductive-parts, including those listed below, to the main earthing terminal:

i. water installation pipes,

ii. gas installation pipes,

iii. other installation pipework and ducting,

iv. central heating and air conditioning systems,

v. exposed metallic structural parts of the building.

Any connection of a lightning protection system to the protective equipotential bonding should be carried out in accordance with *BS EN 62305*.

4.3.3 Automatic disconnection in case of a fault

The impedance of the earth fault loop and the characteristics of the relevant protective device must be co-ordinated such that when an earth fault occurs, the fault current is of sufficient magnitude to operate the circuit protective device, such as a fuse, circuit-breaker or RCD to automatically disconnect the supply to the faulty circuit within a prescribed time.

A protective device such as a fuse, circuit-breaker or RCD is to be provided and the circuit designed such that the device operates and disconnects the circuit in the event of a fault of negligible impedance between a line conductor and an exposed-conductive-part or a protective conductor.

Maximum disconnection times stated in Table 4.1 shall be applied to final circuits with a rated current not exceeding:

i. 63 A with one or more socket-outlets, and

ii. 32 A supplying only fixed connected current-using equipment

Table 4.1 Maximum disconnection times (*Where* 120 V<U_0≤230 V)

	Final circuits with currents not exceeding items i and ii	Final circuits not covered by items i and ii	Distribution circuit
TN System	0.4 s	5 s	5 s
TT System	0.2 s	1 s	1 s

With reference to Table 4.1, where in a TT system disconnection is achieved by an overcurrent protective device and protective equipotential bonding is connected to all the extraneous-conductive-parts within the installation in accordance with Regulation 411.3.1.2, the maximum disconnection times applicable to a TN system may be used.

TN system

The characteristics of the protective devices and the circuit impedances are required to fulfil the following requirement (Regulation 411.4.4 refers): $Z_s \times I_a \leq U_0 \times C_{min}$

Where:

Z_s is the impedance in ohms (Ω), the fault loop comprising:

- the source,

- the line conductor up to the point of the fault, and

- the protective conductor between the point of the fault and the source.

I_a is the current in Amperes (A) causing the automatic operation of the disconnecting device within the time specified in Table 4.1

U_0 is the nominal AC rms or DC line voltage to Earth in volts (V).

C_{min} is the minimum voltage factor to take account of voltage variations - the given value is 0.95.

Table 4.2 indicates where maximum values for earth fault loop impedance for a U_0 of 230 V can be found in *BS 7671*.

Where an RCD is used to provide automatic disconnection in the event of a fault, Regulation 411.4.204 states that the maximum values of earth loop impedance found in Table 41.5 of BS 7671 apply.

Table 4.2 Locations of maximum earth fault loop impedance values within *BS 7671* with respect to required disconnection time.

Required disconnection time (s)	Protective device type	Table Number within the 411.4 Regulation group
0.4	*BS 88-2* fuse systems E and G *BS 88-3* fuse system C *BS 3036* *BS 1362*	41.2
0.4 to 5	*BS EN 60898* circuit-breaker *BS EN 61009* RCBO	41.3
5	*BS 88-2* fuse systems E and G *BS 88-3* fuse system C *BS 3036* *BS 1362*	41.4

TT system

Every exposed-conductive-part shall be connected to a common earth electrode via the main earthing terminal, (Regulation 411.5.1 refers).

If two or more protective devices are in series, the exposed-conductive-parts may be connected to separate earth electrodes corresponding to each protective device. It should be noted that any simultaneously accessible exposed-conductive-parts must be connected to the same earthing system (Regulation 411.3.1.1 refers).

The preferred protective device for fault protection is an RCD (Regulation 411.5.2 refers) but where an RCD is used, as it will be in most cases, overcurrent protection must nonetheless be provided by a fuse or a circuit-breaker, or, alternatively a residual current operated circuit-breaker with integral overcurrent protection (RCBO) may be employed.

Where an RCD is used for fault protection, the following conditions are to be fulfilled for the circuit in question (Regulation 411.5.3 (i) refers):

i. the required disconnection time will be achieved, that is:

- 0.2 s maximum for AC final circuits complying with Regulation 411.3.2.2, or
- 1 s maximum for AC distribution circuits and final circuits complying with Regulation 411.3.2.4, and

ii. $R_A \times I_{\Delta n} \leq 50\ V$

Where:

R_A is the sum of the resistances of the earth electrode and the protective conductor connecting it to the exposed-conductive-parts (in ohms) $I_{\Delta n}$ is the rated residual operating current of the RCD.

Maximum values for earth fault loop impedance for RCDs are given in Table 41.5 of *BS 7671*.

4.3.4 Additional protection by RCD

BS 7671 recognises this measure as reducing the risk of electric shock in the event of the failure of

i. basic protection, and / or

ii. fault protection or

iii. the carelessness of users of the installation.

The measure must not be used as the sole means of protection against contact with live parts (Regulation 415.1.2 refers) and so a protective measure such as ADS must always be provided.

Regulation 410.3.2 states that additional protection by means of an RCD having a rated residual operating current ($I_{\Delta n}$) not exceeding 30 mA is specified as part of a protective measure under certain conditions of external influences, such as:

- where cables are concealed in walls and partitions (see Regulations 522.6.202 and 522.6.203),
- in certain special locations as given in the corresponding Part 7, such as circuits supplying equipment in bathrooms (Section 701).

Where a person makes contact with a live part of a circuit to which this type of protection is provided, and such contact results in a current to Earth of not less than the rated residual operating current of the RCD ($I_{\Delta n}$), the RCD disconnects the supply to the circuit. Note that an RCD cannot detect the current flowing through the body of a person making contact between line and neutral conductors only, and will therefore provide no protection in this circumstance.

The requirements for the rated residual operating current ($I_{\Delta n}$) and operating time of the RCD are given in Regulation 415.1.1. The device is required to have an $I_{\Delta n}$ not exceeding 30 mA and an operating time not exceeding 40 ms at a residual current of 5 $I_{\Delta n}$.

With certain exceptions, additional protection by the use of an RCD is required to be provided for every socket-outlet not exceeding a rating of 32 A and for mobile equipment not exceeding a rating of 32 A for use outdoors (Regulation 411.3.3 refers).

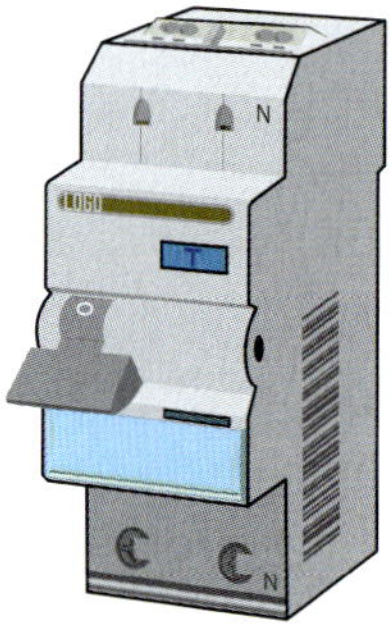

Fig 4.4 Typical RCD used to provide additional protection

4.4 Protective measure: Double or reinforced insulation

Double or reinforced insulation is a protective measure in which either basic protection is provided by basic insulation, and fault protection is provided by supplementary insulation, or both basic and fault protection are provided by reinforced insulation between live parts and accessible parts (Regulation 412.1.1 refers).

Double or reinforced insulation is intended to prevent the appearance of a dangerous voltage on the accessible parts of electrical equipment through a fault in the basic insulation. There is no provision for the connection of exposed metalwork of the equipment to a protective conductor, and there is no reliance upon the earthing arrangements in the fixed wiring of the installation.

The protective measure of double or reinforced insulation is generally applicable in all situations.

In the case of an installation in domestic premises, double or reinforced insulation will not be the main protective measure and will be applied in addition to ADS for a circuit or part thereof, or for an item of current-using equipment such as a luminaire.

Indeed its use as the sole protective measure in premises, such as domestic installations, not under effective supervision is precluded by Regulation 412.1.2.

It is likely that most equipment providing this protective measure will comprise type-tested items bearing the symbol shown in Fig 4.5.

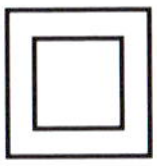

Fig 4.5 Class II equipment should be identified by its construction mark.

A wiring system is considered to meet the requirements of Regulation Group 412.2 for basic and fault protection if:

- the rated voltage of the cable is not less than the nominal system voltage and at least 300 / 500 V, and
- adequate mechanical protection of the basic insulation of the cable has been provided by one or more of the following:
- the non-metallic sheath of the cable, or
- non-metallic trunking or ducting or non-metallic conduit (see Regulation 412.2.4.1).

4.5 Protective measure: Electrical separation

There are two main principles underlying protection by electrical separation:

- neither the source of the supply,
- nor any live parts of the separated circuit should be connected to any other circuit or to Earth.

Thus, in the event of a single fault to an exposed-conductive-part of equipment in the separated circuit, fault protection is afforded because there is no path for fault current to return to the source.

Electrical separation is a protective measure in which:

- basic protection is provided by insulation of live parts or by barriers or enclosures, and
- simple separation of the separated circuit from other circuits and from Earth provides fault protection. That is, basic insulation provides the required separation between circuits or between a circuit and Earth. (Regulation 413.1.1 refers).

Unless the particular requirements of Regulation Group 418.3 are met with regard to factors such as provision of warning notices, basic and fault protection, and earthing and bonding arrangements, this protective measure is limited to the supply of one item of current-using equipment supplied from one unearthed source with simple separation.

For an installation in a dwelling, the only likely application of the use of electrical separation is a shaver supply unit complying with *BS EN 61558-2-5*.

Shaver supply units to *BS 3535* may still be found in older installations and may remain in service subject to their remaining in a sound and serviceable condition.

Because electrical separation has been utilised as the protective measure, there can be no potential difference across that person. Where a shaver supply unit incorporates a metal plate, this normally is required to be earthed.

4.6 Protective measure: Extra-low voltage provided by SELV or PELV

Section 414 of *BS 7671* contains the requirements for extra-low voltage provided by SELV or PELV, which are summarised below.

A SELV system is an extra-low voltage system electrically separated from Earth and from other systems. To this end the live parts of a SELV system must have:

- Basic insulation between live parts and other SELV or PELV circuits,
- Basic insulation between live parts and Earth, and
- Protective separation from live parts of any other systems by means of double or reinforced insulation, or basic insulation and protective screening for the highest voltage present. (Regulation 414.4.1 refers).

A PELV system meets all the requirements for a SELV system except those relating to electrical separation from Earth.

Protection by extra-low voltage provided by SELV or PELV requires all of the following.

- Limitation of voltage in the SELV or PELV system to the upper limit of voltage Band I (50 V AC or 120 V DC).

- Protective separation of the SELV or PELV system from all circuits other than SELV and PELV circuits.
- Basic insulation between the SELV or PELV system and other SELV or PELV systems.
- For SELV systems only, basic insulation between the SELV system and Earth.

Extra-low voltage is generally considered insufficient to present a hazard of electric shock (as defined) in dry situations where the person protected has a body resistance within normal limits. In certain locations, including those containing a bath or shower, the requirements of Part 7 limit the value of the extra-low voltage to lower than 50 V AC or 120 V DC.

If the nominal voltage of a SELV or PELV circuit exceeds 25 V AC or 60 V DC, or if the equipment will be immersed, basic protection must be provided by insulation, or barriers or enclosures.

4.6.1 Requirements for basic protection and fault protection

Basic protection and fault protection are provided subject to all of the listed conditions being met.

- The nominal voltage cannot exceed the upper limit of voltage Band I .
- The supply is from a permitted source (see Section 4.6.2 of this Guide).
- The conditions of Regulation 414.4 are met.

If the system is supplied from a higher voltage system by equipment which provides at least simple separation between that system and the extra-low voltage system but which does not meet the requirements for a SELV and PELV source, the requirements for FELV may be applicable. In such cases, the system does not meet the requirements for SELV or PELV and therefore must not be used in an application where use of these systems is required.

4.6.2 Sources for SELV and PELV

The following sources may be used for SELV and PELV systems.

- A safety isolating transformer in accordance with *BS EN 61558-2-6*.
- A source of current providing a degree of safety equivalent to that of the safety isolating transformer such as a motor-generator with windings providing equivalent isolation.
- An electrochemical source (for example a battery) or another source independent of a higher voltage circuit (such as a diesel-driven generator).
- Certain electronic devices complying with appropriate standards, where provisions have been taken to ensure that, even in the case of an internal fault, the voltage at the

outgoing terminals cannot exceed 50 V AC or 120 V DC. Higher voltages at the outgoing terminals are, however, permitted if it is ensured that, in case of contact with a live part or in the event of a fault between a live part and an exposed-conductive-part, the voltage at the output terminals is immediately reduced to 50 V AC or 120 V DC or less.

- A mobile source supplied at low voltage, for example a safety isolating transformer or a motor-generator, shall be selected and erected in accordance with the requirements for protection by the use of double or reinforced insulation.

Sources for the supply of SELV and PELV circuits must comply with the above requirements and this is to ensure that there is no possibility of a dangerous voltage occurring in a SELV or PELV circuit due to an internal fault in the source. For example, an insulation breakdown in a step-down transformer could raise the voltage of the secondary circuit to a dangerous level.

Such danger is prevented by using a safety isolating transformer complying with *BS EN 61558-2-6*, or one of the other sources, listed previously in this section. Fig 4.7 shows a suitable transformer together with circuit configuration and identifying symbol.

SELV and PELV circuit wiring is electrically (and often physically) separated from the wiring of higher voltage systems. This is to minimise the possibility of introducing a higher voltage into a SELV or PELV circuit in the event of an insulation failure in a conductor of a higher voltage circuit.

To further reduce the risk of electric shock, conductors and exposed-conductive-parts of a SELV system are not connected to Earth or to other systems.

BS 7671 requires further protection in locations of increased shock risk, including bathrooms, rooms containing a shower, and swimming pools. Such protection may include limiting the nominal voltage to a value not exceeding 25 V AC rms or 60 V ripple-free DC (or in some cases an even lower limit), and the provision of insulation, barriers or enclosures for protection against contact with live parts.

In some locations of increased shock risk, such as certain zones of swimming pools, SELV is the only protective measure against electric shock permitted by *BS 7671*.

The wiring systems of SELV or PELV circuits are required to have protective separation from the live parts of other circuits, which must have at least basic insulation, and this separation must be achieved by one of the following arrangements.

- SELV and PELV circuit conductors enclosed in a non-metallic sheath or insulating enclosure in addition to basic insulation.
- SELV and PELV circuit conductors separated from conductors of circuits at voltages higher than Band I by an earthed metallic sheath or earthed metallic screen.

- Circuit conductors at voltages higher than Band I may be contained in a multi-conductor cable or other grouping of conductors if the SELV and PELV conductors are insulated for the highest voltage present.
- The wiring systems of other circuits are in compliance with the requirements of Regulation 412.2.4.1 for double or reinforced insulation.
- Physical separation.

4.6.3 Accessories

A socket-outlet or luminaire supporting coupler in a SELV or PELV circuit must be dimensionally incompatible with those used for any other system in use in the same premises (see Regulation 414.4.3). Additionally Regulation 414.4.4 states that plugs and socket-outlets in a SELV system must not have a protective conductor contact.

4.6.4 Additional requirements for SELV circuits.

Exposed-conductive-parts of a SELV circuit must not be connected to Earth, or to protective conductors or to exposed-conductive-parts of another circuit.

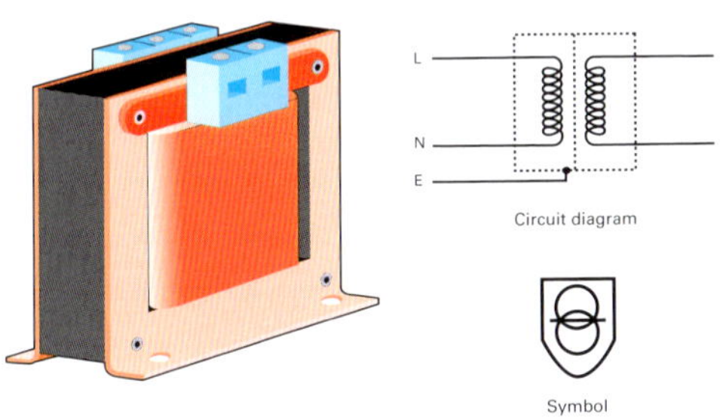

Fig 4.7 Safety isolating transformer complying with *BS EN 61558-2-6*

If exposed-conductive-parts of a SELV circuit could come into contact, either fortuitously or intentionally, with the exposed-conductive-parts of other circuits, the requirements of the protective measure SELV will not have been met and protection against electric shock will no longer depend solely on protection by SELV, but also on the protective measures applied to the circuit to which the latter exposed-conductive-parts are related.

4.7 Additional protection

Additional protection is required, in many cases, to supplement the selected protective measure under certain conditions of external influences and in certain special locations (see Part 7 of *BS 7671*). The additional protection required may be either provided by:

- an RCD, or
- supplementary equipotential bonding.

4.7.1 RCDs

RCDs having a rated residual operating current not exceeding 30 mA can be installed to provide additional protection in the event of failure of the provision for basic protection and / or the provision for fault protection or the carelessness of users in AC systems. (Regulation 415.1.1 refers).

However, the use of such devices is not recognised as a sole means of protection and does not obviate the need to apply one of the recognised protective measures (Regulation 415.1.2 refers).

4.7.2 Supplementary equipotential bonding.

Where supplementary equipotential bonding is installed it must include all simultaneously accessible exposed-conductive-parts of fixed equipment and extraneous-conductive-parts. Such bonding must be connected to the protective conductors of all equipment of the installation or part thereof or location including those connected to socket-outlets where installed.

The resistance *R* between simultaneously accessible exposed-conductive-parts and extraneous-conductive-parts should fulfil the following condition in an AC system:

$R \leq \dfrac{50\ \text{V}}{I_a}$ Where I_a is the operating current in Amperes of the protective device.

If the protective device is an RCD, I_a is the rated residual operating current ($I_{\Delta n}$). If the protective device is a fuse or circuit-breaker, I_a is the current causing operation of the protective device within the specified time (Regulation 415.2.2 refers).

CHAPTER 5
CONSUMER UNIT AND AND DISTRIBUTION BOARD ARRANGEMENTS

5.1 Introduction

This Chapter explains the requirements of *BS 7671* affecting the choice of consumer unit.

A consumer unit is a particular type of distribution board designed for use in domestic and similar installations. A consumer unit used for domestic use and intended to be operated by ordinary persons should meet the requirements of *BS EN 61439-3: 2012 (as amended) Low-voltage switchgear and controlgear assemblies. Part 3: Distribution boards intended to be operated by ordinary persons (DBO)*

In its simplest form, a consumer unit comprises:

- an enclosure,
- a main switch that, when operated, disconnects both live (that is line and neutral) conductors (Regulation 462.1.201 refers),
- a number of protective devices such as fuses, circuit-breakers and RCDs,
- neutral and earth termination bars,
- interconnecting busbars / wiring.

However, and with reference to the definition of a consumer unit given in Part 2 of *BS 7671*, a consumer unit may also contain signalling devices, transformer units and the like. It should be noted that any component to be so installed should be type-tested as suitable for use within the particular make and model of consumer unit.

Many consumer units in older installations forming part of a TN system do not contain RCDs. In such installations where supplementary protection has been provided, the RCD typically is installed some way downstream of the origin of the final circuits, for example to protect individual socket-outlets likely to be used to supply portable equipment for use outdoors.

In the case of installations forming part of a TT system, consumer units having an RCD acting as the main switch are not uncommon. Such an arrangement would be unlikely to satisfy the requirement of Regulation 314.1, to minimise inconvenience in the event of a fault in anything other than the simplest of installations.

Starting with a basic enclosure and main switch, a consumer unit may be arranged with:

- each final circuit protected individually by an RCBO (see Section 3.5.2 of this Guide) and allowing the possibility of circuit-breaker or fuse protection for those circuits where RCD protection is not appropriate,
- a 'split' unit containing two or more 2-pole RCCBs, each protecting a number of final circuit ways, which have circuit-breakers or fuses fitted to provide overcurrent protection,

- a 'split' unit containing two or more 2-pole RCCBs as described above but with a number of final circuit ways not protected by the RCCBs thus allowing the possibility of fault protection by a circuit-breaker or fuse for those circuits where RCD protection is not required.

5.2 Split-load arrangements

Split-load consumer units can provide a range of circuit configurations. Where it is not necessary to provide individual RCD protection for each final circuit, a suitable split-load consumer unit enables those circuits requiring RCD protection to be provided with such protection as a group, whilst also making provision for any circuits that do not require RCD protection.

Fig 5.1 shows one such arrangement for a consumer unit for use in an installation forming a part of a TN system in which the main switch supplies two RCCBs. The final circuits of the installation are shared between the two RCCBs in a manner that would minimise disruption and avoid danger in the event of a fault.

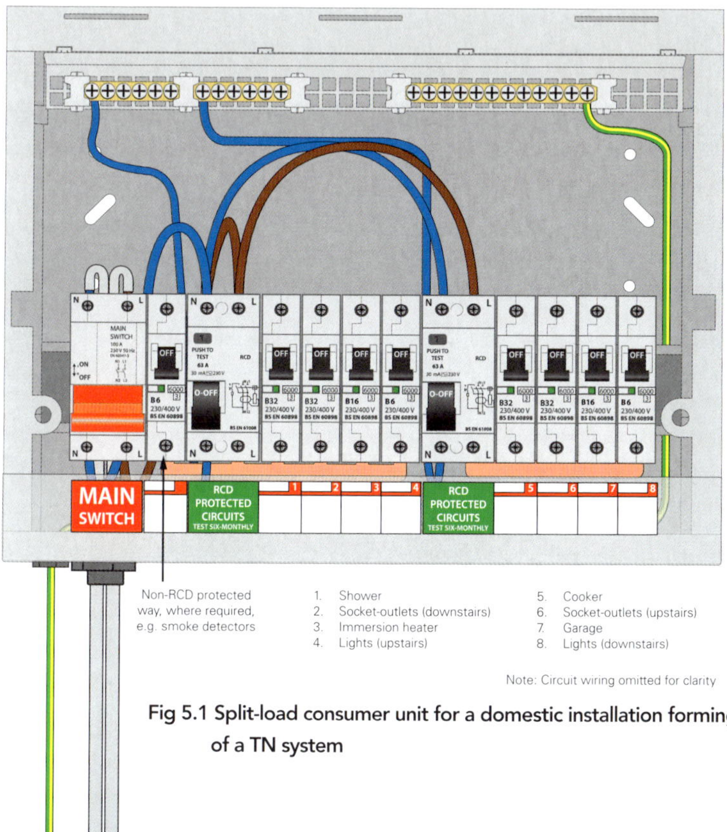

Fig 5.1 Split-load consumer unit for a domestic installation forming part of a TN system

Fig 5.1 shows one circuit that is not protected by an RCCB. Such an arrangement might be appropriate for a fire detector / alarm unit where the circuit has been run on the surface, or if concealed within a wall, the cable:

- incorporates an earthed metallic covering, or
- has been installed in earthed conduit, trunking or ducting, or
- otherwise sufficiently mechanically protected to prevent penetration by nails, screws or similar

It should be noted that it remains acceptable to connect fire detector / alarm units of a Grade D system to a separately electrically protected and regularly used local lighting circuit using 'soft skinned' cables such as thermoplastic insulated / sheathed cable typically used to wire the lighting and other circuits within the dwelling (Clause 15.5 a) ii) of *BS 5839-6: 2013* refers).

Whilst, preferably, a circuit supplying fire detector / alarm units should not be protected by an RCCB, such protection should be provided where it is necessary to do so in order to meet the requirements of *BS 7671*, such as in installations forming a part of a TT system or where cables are installed in a wall at a depth of less than 50 mm.

5.3 Advantages and disadvantages of a split-load consumer unit

A split-load consumer unit has the major advantage that it can provide a convenient means of meeting the requirements of *BS 7671* for most dwellings.

RCD protection of a group of final circuits has the disadvantage that an earth fault sufficient to operate the RCCB protecting any one circuit of the group will cause all the circuits of that group to be disconnected.

The installation designer needs to decide whether the use of a split-load consumer unit for a particular application meets with the requirements of Regulations 314.1 and 314.2. Regulation 314.1 requires every installation to be divided into circuits as necessary to avoid danger and minimise inconvenience in the event of a fault, and to facilitate safe operation, inspection, testing and maintenance. Regulation 314.2 requires, amongst other things, that due account shall be taken of the consequences of the operation of any single protective device).

5.4 Series connected RCDs

A configuration employing a split-load consumer unit with RCDs connected in series is illustrated in Fig 5.2. Such a configuration might be found in an installation forming part of a TT system. Each individual circuit would be protected by its own overcurrent protective device.

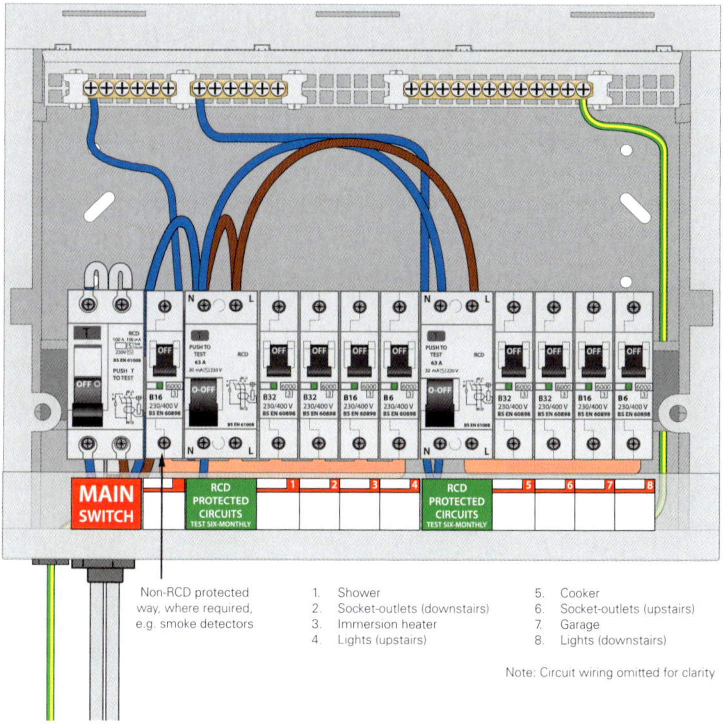

Fig 5.2 Simplified diagram of a split-load consumer unit with time-delay RCCB main-switch for use in a domestic installation forming part of a TT system

Fig 5.3 shows a consumer unit having a main switch and a number of circuits 'ways' protected individually by RCBOs. Where RCCB protection is not appropriate (as discussed previously in Section 5.2) a particular circuit could be provided with fault protection by a circuit-breaker or fuse.

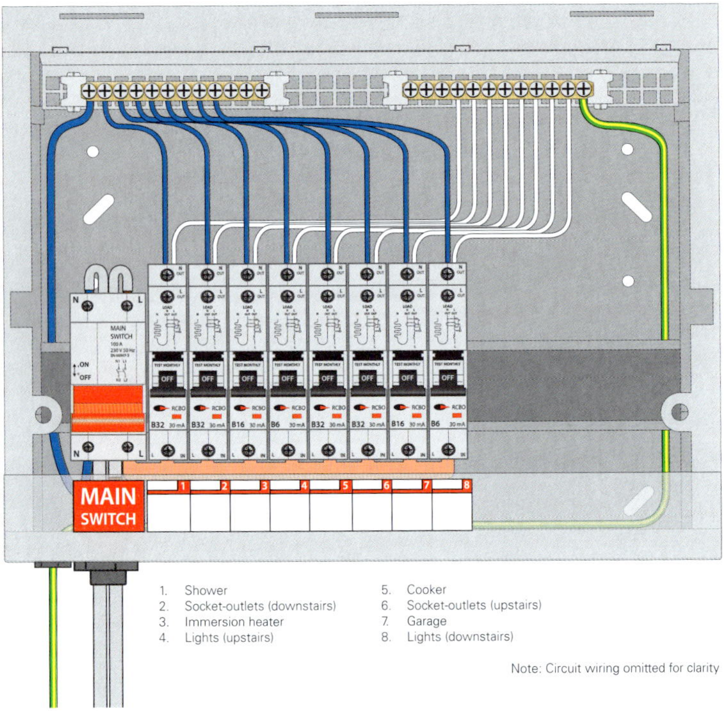

Fig 5.3 Consumer unit suitable for use with circuit-breakers, fuses and RCBOs

5.5 Selectivity between RCDs in series

On occasion, a domestic electrical installation may include RCDs connected in series, particularly where the installation forms part of a TT system. Where such RCDs provide fault protection or otherwise prevent danger, and where selectivity in their operation is necessary to prevent danger, Regulation 536.4.1.4 of *BS 7671* requires that their characteristics are such that selectivity is achieved.

The need for selectivity is also implicit in Regulation 314.2. This requires that a separate circuit be provided for each part of an installation that needs to be separately controlled to prevent danger.

Selectivity between RCDs connected in series is also required for reasons other than to prevent danger, including to minimise inconvenience in the event of a fault.

Even where it is not strictly necessary to provide selectivity between RCDs to prevent danger or to minimise inconvenience, it is considered good practice to do so, not the least because it can greatly simplify identification and / or location of the device that has operated.

Where two or more RCDs are connected in series, selectivity will generally be achieved by selecting time-delayed characteristic(s) for the upstream device(s) and having a rated residual operating current of the upstream device of at least three times the rating of the downstream device (Regulation 536.4.1.4 refers).

Selectivity between RCDs can really only be accomplished by delaying the tripping of the RCD upstream such that the downstream RCD has time to disconnect the faulty circuit before the time delay of the upstream RCD has expired.

It should be noted that RCDs used for additional protection are not permitted to have an operating time delay. Such RCDs must, as stated in Regulation 415.1.1, have a rated residual operating current not exceeding 30 mA.

Furthermore, where RCDs with a time delay are used to provide fault protection, the total operating time is not to exceed the applicable maximum disconnection times required by Chapter 41 of *BS 7671*.

British Standards *BS EN 61008-1* and *BS EN 61009-1* for residual current operated circuit-breakers with and without integral overcurrent protection, respectively, classify residual current operated circuit-breakers with time-delay as 'type S' (for selectivity).

5.6 Selectivity between an upstream overcurrent protective device and an RCD

In the event of an earth fault occurring, the resultant current may exceed the instantaneous tripping current of an overcurrent protective device (OCPD) upstream of an RCD causing the upstream device to operate first. This situation is more likely for devices having low rated currents (I_n) and, in the case of circuit-breakers, having a type B time / current characteristic.

In such cases, where selectivity between an upstream OCPD and downstream RCDs is required, the downstream device should be an RCBO or CBR (circuit-breaker incorporating residual current protection) and the requirements for selectivity according to Regulation 536.4.1.5.1 should be applied (see Fig 5.4).

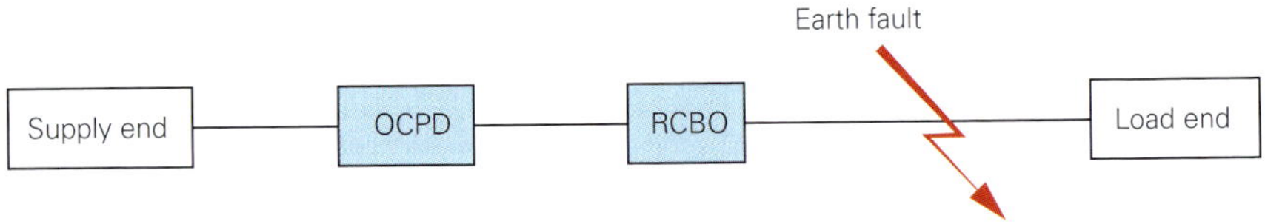

Fig 5.4 Selectivity between an upstream overcurrent protective device and an RCBO

5.7 Selectivity between an RCD and a downstream overcurrent protective device

In the event of an earth fault occurring, the resultant current may be lower than the instantaneous tripping current of the downstream OCPD causing the upstream RCD to operate first.

In such cases, where selectivity between an upstream RCD and a downstream OCPD is required, the downstream device should be an RCBO and the requirements for selectivity according to Regulation 536.4.1.4 should be applied (Regulation 536.4.1.6 refers). Where such an arrangement is installed, the upstream RCD will need to be of the type S or time-delay type as described earlier in this Chapter.

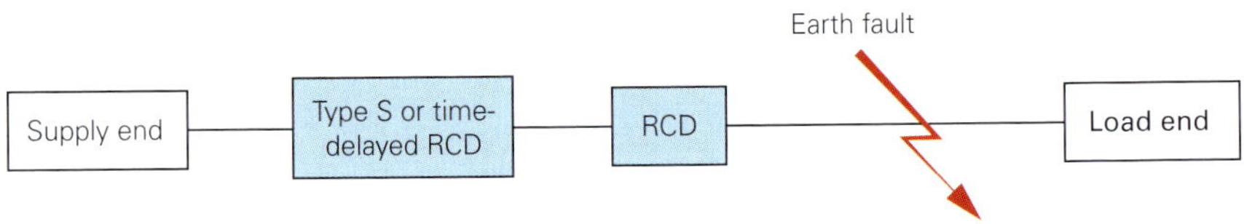

Fig 5.5 Selectivity between an upstream RCD and an overcurrent protective device

Total selectivity is not always required

It should be borne in mind that such selectivity between upstream and downstream devices is not essential in all cases. Partial selectivity between devices is acceptable where this does not adversely affect the safety of the installation (Regulation 536.3 refers).

To clarify using the example of installations in typical domestic premises, it is acceptable to 'split' the circuits connected to a consumer unit over, say, two RCDs; each of which protects a number of final circuits (see Fig 5.1 and 5.2). It should of course be understood by the designer, and made clear to the customer that operation of one of the RCDs in such a 'split board' arrangement would cause the loss of all circuits downstream of it.

5.8 Installations forming part of a TT system

Irrespective of the type of material the consumer unit is manufactured from, where an installation forms part of a TT earthing system, Regulation 411.5.2 typically requires the use of one or more RCDs as a protective measure for fault protection, and they shall be placed at the origin of the installation (see also Regulation 411.5.3).

Whilst it is not possible to place a double-pole RCD at a position that will eliminate all risk associated with protecting meter tails from an earth fault, it is possible to reduce the risk of a fault to earth by positioning the consumer unit in close proximity to the energy meter.

Although metal-clad consumer units are demonstrably non-combustible when compared with equivalent all-insulated units, metal-clad units have the disadvantage that if they are installed in an installation forming part of a TT system, an earth fault occurring between the incoming meter tails and the metal case of the consumer unit will result in the metal case becoming live.

Such a fault will cause all other conductive parts connected to the main earthing terminal (MET) to rise to a potential approaching close to that of the single phase supply to Earth. Under such circumstances, due to the relatively high external earth fault loop impedance of a typical TT system, the service fuse is unlikely to operate and even when it does it may not operate within the 1 s disconnection time required by *BS 7671* (Regulation 411.3.2.4 refers)

However, by adopting longstanding industry practice, metal-cased consumer units have been used safely in installations forming part of a TT system for many years (refer Fig 5.6).

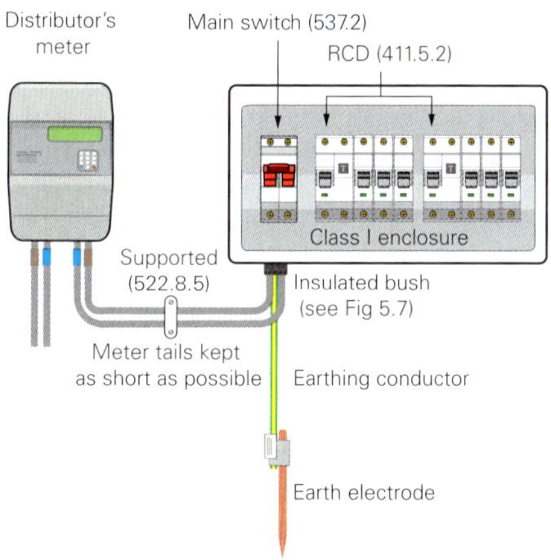

Fig 5.6 Effect of fault to earth between incoming tails and the metal case of a consumer unit

The approach generally adopted, is to house the RCD within the consumer unit and to use thermoplastic or thermosetting insulated and sheathed cable for the meter tails and fit an insulated bush, located where the meter tails enter the metallic consumer unit or protective cabinet.

The meter tails entering the consumer unit must have double or reinforced insulation (Regulation 531.3.5.3.2.201 refers).

An exploded view of a typical insulated bush is shown in Fig 5.7.

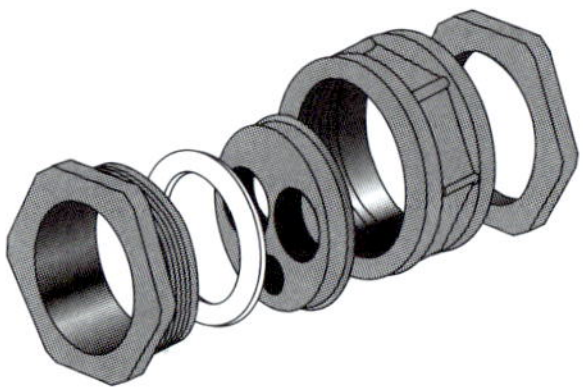

Fig 5.7 Insulated bushes for entries to Class I enclosures

Cable entries into the consumer unit must maintain a degree of protection of at least IPXXD or IP4X for the top surface (Regulation 416.2.2 refers) and IPXXB or IP2X elsewhere (Regulation 416.2.1 refers).

To prevent any potential contact with the meter tails and the metal consumer unit and / or enclosure, the installer should ensure that the conductors are correctly installed with no appreciable mechanical strain on the connections (Regulations 522.8.5, 526.6 and 526.7 refer), and the connections are tightened to a torque setting recommended by the manufacturer.

5.9 Protective conductor currents

The cumulative addition of protective conductor currents (commonly referred to as 'earth leakage current') during normal operation in the individual circuits of a group of circuits protected by an RCD, can be sufficient to cause an RCD to operate; for example, where suppressors or filters such as capacitors are connected between line, neutral and circuit protective conductors. In such cases, the installation designer will need to decide whether the use of a split-load consumer unit for a particular application meets with the requirements of Regulation 531.3.2 (i), which requires an RCD to be so selected and the electrical circuits so subdivided such that any protective conductor current likely to occur during normal operation of the connected load(s) does not cause unnecessary tripping of the device.

Information technology and other equipment that exhibits significant protective conductor currents are subject to the requirements of Regulation Group 543.7, the requirements of which are summarised in Section 2.10 of this Guide.

5.10 Consumer unit location

The chosen location for a consumer unit must provide reasonable access to facilitate the operating of switches and protective devices, and to permit inspection, maintenance and access to all connections. Where a consumer unit is mounted in a cabinet, compartment or similar, this should be done in a way that does not impair access.

Switches and protective devices housed in a consumer unit need to be accessible to the user. It should not be necessary, for example, to make use of a ladder or other temporary platform to gain access. Where the user is likely to be elderly or infirm, access to switches and protective devices should comply with *BS 8300: 2009 + A1: 2010 - Design of buildings and their approaches to meet the needs of disabled people – Code of practice.*

5.11 Separation of electrical cables and accessories from the gas installation pipework

Where it is unavoidable to mount a consumer unit or run cables and / or mount equipment in close proximity to pipework forming part of the gas installation (see Fig 5.8), the following shall apply:

- the electrical equipment or cables shall be contained within an insulating enclosure, separated by a dividing barrier or be contained within conduit or trunking; or
- consumer units, main isolating switches and the like shall be mounted at least 150 mm away from any gas pipework; or
- any electrical supply or distribution / circuit cables and any switches, socket-outlets and such shall be separated by a distance not less than 25 mm from any gas pipework or associated equipment.

The mounting of any electrical accessory shall be such that it will not be prevented from operating in normal service from the close proximity of any gas installation pipework (Clause 8.4.2 of *BS 6891* refers).

Where the spacing's shown in Fig 5.8 is unachievable, the gas installation pipework should either have an outer sheathing of electrical insulating material rated to at least 230 V AC, or a panel of electrical insulating material be placed between the electrical and gas installation equipment.

The above details and mounting distances are quoted from *BS 6891: 2015 Specification for the installation and maintenance of low pressure gas installation pipework of up to 35 mm (R11 / 4) on premises.*

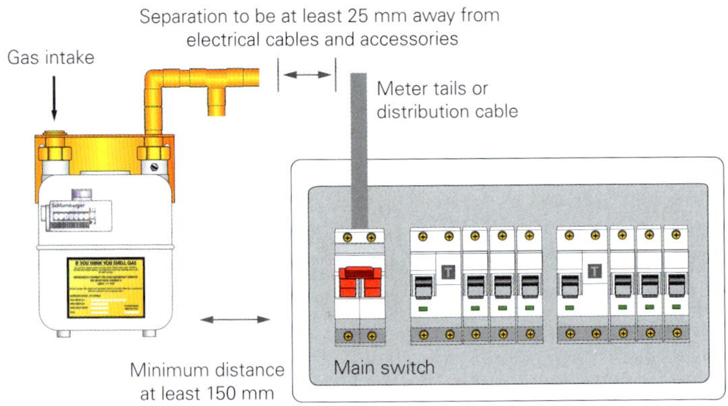

Fig 5.8 Separation distances between electrical and gas services

5.12 Mounting height

Part M of the Building Regulations, for England and Wales, requires reasonable provision to be made for people to gain access and to use the building and its facilities.

Guidance on meeting the requirements of Part M is given in Approved Document M.

Section 8 of Approved Document M, which applies to new dwellings, includes the objective of assisting people whose reach is limited to use the dwelling more easily.

The document indicates that wall-mounted switches, socket-outlets and control devices should be installed at a height between 450 mm and 1200 mm above finished floor level, the dimensions taken from the bottom and top edges of the equipment respectively. Similar recommendations are given for compliance with the building regulations in Scotland and in Northern Ireland by Technical Handbooks and Technical Booklets respectively.

5.13 Dwellings liable to flooding

Guidance on installations where there is a risk of flooding is given in a publication entitled Preparing for floods, and a free electronic copy can be downloaded by visiting the Department of Communities and Local Government's website, at **www.communities.gov.uk**

When undertaking work at a dwelling at risk of flooding, items such as consumer units and energy meters (subject to agreement with the distributor) should be mounted above the expected flood level if practicable.

CHAPTER 6
ISOLATION AND SWITCHING

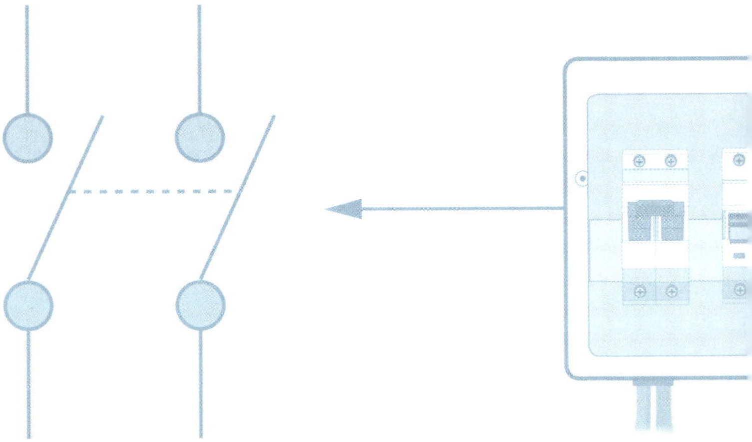

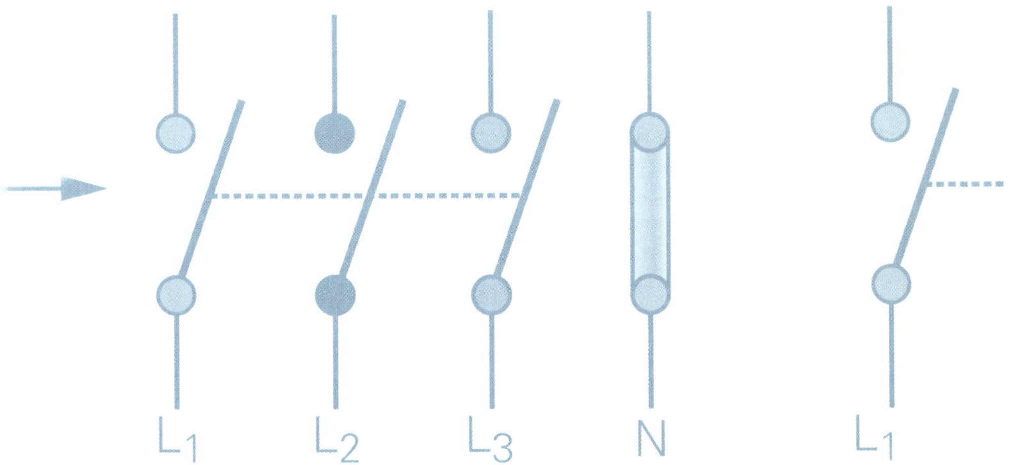

6.1 Introduction and terminology

Section 537 of *BS 7671* details the requirements for isolation and switching that should be provided for an electrical installation and the electrical equipment or machinery therein as appropriate to the particular circumstances.

The term 'isolation and switching' refers to four distinct functions, namely:

- isolation,
- switching off for mechanical maintenance,
- emergency switching, and
- functional switching.

The purpose of isolation (as defined) is to provide protection for a person engaged in work on, or near to, electrical equipment. Once equipment has been securely isolated from the source of energy and any electrical energy has been discharged, persons should be able to safely access parts that would otherwise be live, or may become live, without the risk of danger from electric shock, electric arcing and / or explosion, or electric sparking (where this may cause fire and / or explosion) or from movement of electrically powered equipment and machines.

Every circuit must be capable of being isolated from each of the live supply conductors (Regulation 462.2 refers): a neutral conductor falls within the definition of a live conductor. With permitted exceptions for TN-S and TN-C-S systems, detailed below, it should be possible to isolate the neutral conductor of every circuit from the corresponding supply conductor (see Regulations 462.1.201 and 461.2).

A means of isolating and switching the supply on load must be provided as near as practicable to the origin of every installation. The means of isolation and switching is required to be a main linked switch or a linked circuit-breaker (Regulation 462.1.201 refers).

Isolation of the neutral conductor of a circuit in TN-S or TN-C-S systems is not generally required where that conductor can be reliably regarded as being connected to Earth by suitably low impedance. (Regulation 461.2 refers).

A group of circuits may be isolated by a common means (Regulation 462.2 refers). However, where such an arrangement is being considered, the requirements of Regulation Group 314 relating to division of an installation must be taken into account.

A common example of a single device used to isolate a group of circuits is where the main switch of a consumer unit is operated (switched off), thereby isolating all of the outgoing circuits.

The requirements for the isolation of live supply conductors at the origin, is illustrated in Figure 6.1

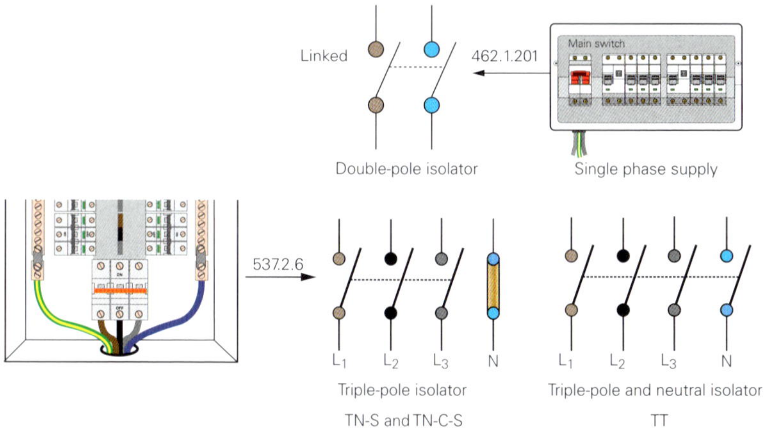

Fig 6.1 Isolation arrangement at the origin of an installation

In most installations, a circuit is formed at the origin consisting of 'meter tails' connecting to the consumer unit. For example, a single-phase domestic supply circuit may consist of either a *BS 1361* or *BS 88-3* fuse (in the service cut-out) connected by the meter tails to a consumer unit as illustrated in Figure 6.2.

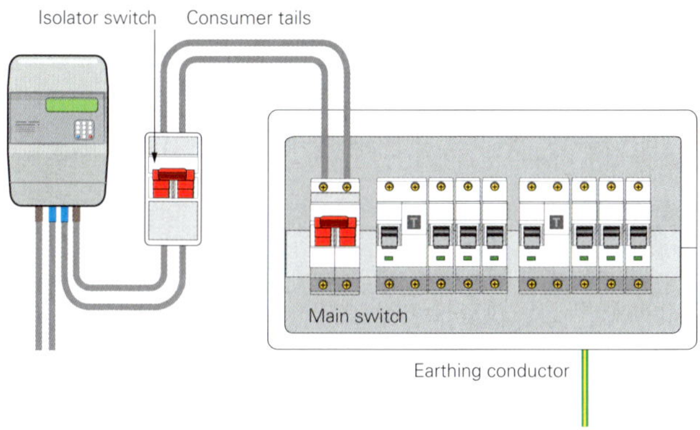

Fig 6.2 Means of isolation for meter tails

Where a device used for isolation has been inserted in series between the energy meter and the consumer unit, Regulation 461.1 states that such a device provided for isolation or switching shall comply with the relevant requirements of Chapter 53 of *BS 7671*.

Persons working in domestic and similar premises should not attempt to access or withdraw the cut-out fuse(s) under any circumstances unless authorised to do so by the relevant distributor.

6.2 Installations supplied from more than one source

Where an installation is supplied from more than one source, a main switch (a means of isolation and switching the supply on load) must be provided for each source of supply. Such situations may arise in dwellings that have small-scale embedded generators, such as solar photovoltaic systems and domestic combined heat and power units. Additionally, where an installation has more than one supply, Regulation 514.15.1 requires a durable warning notice to be permanently fixed and so placed as to warn any person seeking to isolate the installation of the need to operate the isolation and switching device for each source of supply.

Any device used for isolation must be non-automatic (Regulation 537.1 refers) - that is, requiring manual operation. Further, any device used for isolation should be suitably positioned or durably marked such that the installation or the circuit that it isolates is clearly identifiable (Regulation 537.1.2 refers).

6.3 Remote isolation

Where an isolating device for a particular circuit is located remotely from the location in which work activities will be carried out, provision must be made such that the means of isolation can be secured in the open, in the case of off-load isolating devices, the closed position (Regulations 537.2.4 and 537.2.5 refer).

The securing arrangement and the procedures put in place for its release must be such that closing of the means of isolation is not possible until it is safe to do so.

Where the provision for securing in the open position takes the form of a lock or a removable handle, the key or handle is to be non-interchangeable with any others used for similar purposes within the installation.

6.4 Isolating devices

The particular requirements of *BS 7671* applicable to the devices providing isolation are summarised in Table 6.1 and further discussed in this Chapter.

Table 6.1 Summary of requirements for isolation

Requirement	Regulation
Device to isolate all live supply conductors (with certain permitted exceptions)	462.1.201; 537.2.6
A multi-pole device is preferable where multi-pole isolation is needed	537.2.6
Manual operation	537.1
The disconnection and separation is to be secure and the device is unlikely to intentionally re-close	537.2.4; 462.3
Installations having more than one supply	537.1.2
Permitted devices	Table 537.4; 537.2.3
Unsuitable devices	Table 537.4; 537.2.2
Marking	537.2.1; 537.2.7
Discharge of stored energy	462.4
Provision for disconnection of the neutral conductor	461.2; 537.2.8

Devices for isolation, which must be suitable for manual (non-automatic) operation, are required to isolate all live supply conductors from the circuit concerned, except in TN-S or TN-C-S systems, where the neutral conductor need not be isolated or switched if it can be reliably regarded as being connected to Earth by suitably low impedance. This exception does not apply to the main switch of household and similar premises where the means of isolation may be used by ordinary persons (as defined) (Regulation 462.1.201 refers).

Where multi-pole isolation is required, it is preferable for the means of isolation to be a multi-pole isolating device. However, single-pole devices such as fuses or links situated adjacent to each other are permitted (Regulation 537.2.6 refers).

A device for isolation must be selected and / or installed in such a way as to prevent unintentional closure such as may be caused by mechanical shock, accidental knocks or vibration (Regulation 537.2.4 refers).

Each device used for isolation shall be clearly identified by position or durable marking to indicate the installation or circuit it isolates (Regulation 537.2.7 refers).

A plug and socket-outlet or similar device such as a luminaire support coupler may be selected for use as a means of isolation (Regulation 537.1.3 refers).

It should be noted that circuit-breakers, switches and other devices might not meet the requirements for a disconnector and as such may not be suitable as a means to provide isolation. It is important that such devices are not used for isolation purposes unless they meet the requirements of a disconnector as defined in Part 2 of *BS 7671*.

Table 537.4 of *BS 7671* provides guidance on the selection of devices in terms of their suitability to perform the functions of isolation, emergency switching and functional switching.

Some product standards, for example *BS EN 60947-3*, require the product to be marked with a symbol to indicate its suitability for isolation. Examples of the symbols used for some of the more common means of isolation are shown in Fig 6.3. The left hand column of the chart shows devices suitable for off-load isolation. The right hand column shows devices suitable for on-load making, breaking and isolating.

Semiconductor devices must not be selected for the purposes of isolation (Regulation 537.2.2 refers).

Table 6.2 Symbols used for some of the more common means of isolation

Isolating	Making, breaking and isolating
Disconnector 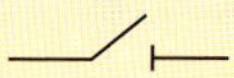	Switch-disconnector 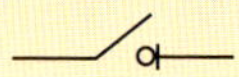
Disconnector-fuse 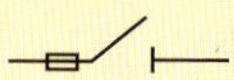	Switch-disconnector-fuse 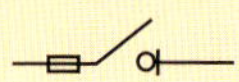
Fuse-disconnector 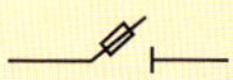	Fuse-switch-disconnector 

6.5 Switching off for mechanical maintenance

Switching off for mechanical maintenance is provided for the benefit of persons who are not necessarily electrically-skilled and may not necessarily be competent to avoid the dangers which electricity may create. Great care must be taken in the design and implementation of measures for switching off for mechanical maintenance to ensure that they provide safe working conditions, particularly where non-electrically skilled persons are likely to be undertaking the maintenance.

6.6 Emergency switching

In domestic installations, there is generally no requirement to install devices such as emergency stop buttons. However, installers should be aware that even though *BS 7671* does not permit *BS 1363* plugs and socket-outlets to be selected as a means of providing emergency switching (Table 537.4 and Regulation 537.3.3.3 refer), in an emergency such as a fire involving deep fat fryers or a tumble dryer, the user may use a *BS 1363* plug and socket-outlet to disconnect the equipment.

When installing socket-outlets that may, in an emergency, be used to disconnect equipment such as that previously described, the installer should ensure that the socket-outlets are positioned so that, when the equipment is in use, the socket-outlets are readily accessible to the user.

6.7 Functional switching

A functional switching device should be provided for each part of a circuit that needs to be controlled independently of other parts of the installation and shall be selected in accordance with Table 537.4.

Regulation 537.3.1.2 states that all items of current-using equipment requiring control should be controlled by a functional switching device that is appropriate in relation to particular circumstances. In the case of appliances, the device for functional switching may be, and often is, an integral part of the appliance.

A single functional switch may be used to control several items of equipment where those items are intended to operate simultaneously, such as a lighting switch configured to control multiple downlighters in a single room.

Where there is possibility of confusion, a functional switch should be labelled to indicate its purpose. For example, where a functional switch controls a water heater in another part of the house or that otherwise might not readily be associated with the switch, the switch should be labelled to indicate its purpose. Where the purpose of a functional switch is clearly evident, a label need not be provided to indicate its purpose.

Functional switching is provided for the users of electrical installations for normal operating purposes to control items of current-using equipment. The equipment may be controlled either individually or in groups.

The functional switching / control of equipment may be either manual or automatic. A time switch is an example of an automatically-operated functional switch.

Examples of functional switches include:

- a switch in a socket-outlet,
- a contactor switching the supply to start and stop an electric motor,
- push buttons used to control a contactor or relay,
- a thermostat used to switch an immersion heater or central heating 'on' and 'off',
- a pressure switch or level switch used to control a pump,
- lighting switches and dimmers.

Functional switching devices are generally required to be capable of switching the load 'ON' and 'OFF'.

All functional switching devices are required by Regulation 537.3.1.2 to be suitable for the most onerous duty intended, and should be selected to have a utilisation category appropriate to the type of load being switched.

Opening contacts are not essential

Functional switching devices do not necessarily have contacts that open and close.

Rheostats, potentiometers, and semiconductor devices that control current electronically, such as a wall mounted dimmer switch, may be used as functional switching devices (Regulation 537.3.1.3 refers).

Disconnectors (off-load isolators), fuses and links are not considered to be suitable devices for functional switching and are not to be used for this purpose (Table 537.4 refers).

A plug and socket-outlet having a rated current not exceeding 32 A may be used for functional switching in AC circuits (Table 537.4 of *BS 7671* refers).

Table 537.4 contains a note advising that circuit-breakers and RCDs are circuit-protective devices and as such they are not intended for frequent load switching. However, for the purposes of isolation or emergency switching the infrequent switching of circuit-breakers is acceptable. Where a more frequent duty is required manufacturer's instructions should be consulted and, where necessary, an alternative device provided.

CHAPTER 7
IDENTIFICATION AND LABELLING

7.1 Introduction and general requirements

Labelling provided for identification, information or warning purposes should be clearly legible and durably marked. Any inscription on an item of equipment or a label should remain legible for the lifetime of the installation. For example, inscriptions might be engraved or embossed into suitable material, although other suitable methods of marking are not precluded.

In general, labels must be sized such that they are clearly visible from the normal viewing position. Any labels affixed to, or adjacent to, an item of equipment should be permanent.

Many labels may have black letters on a white background, but where identification of a device for isolation and switching forms part of a safety sign, or provides clarifying information supportive of a safety sign, then the label should comply with, and be colour-coded in accordance with, *BS ISO 3864-1:2011 Graphical symbols. Safety colours and safety signs: Design principles for safety signs and safety markings.*

For example, notices warning of the need to operate appropriate isolating devices and identifying their location should have black lettering on a yellow background. Fire equipment signs should have white lettering, and where necessary white directional arrows, on a red background. Additionally, the requirements of the *Health and Safety (Safety Signs and Signals) Regulations 1996 must be met, where applicable.*

Labels should be fixed to equipment such as switchgear or controlgear by suitably secure means such as screws, rivets or adhesive. The means of fixing a label to an item of equipment should always be selected such that the integrity of the ingress protection of the enclosure is not compromised.

Consideration should always be given to the effects of environmental conditions such as heat, water, pollution or corrosive substances, when determining whether a label may be considered to be durably marked and permanent.

The following are examples of informative labels required by *BS 7671*.

- Periodic inspection and testing (date of last inspection and recommended date of next inspection)
- Identification of switchgear and controlgear
- Location of means of isolation and switching
- Identification of source(s) of supply
- Identification of circuit protective devices
- Identification of current-using equipment
- Warning of dual supply where an installation includes a generating set such as a solar photovoltaic (PV) system or a wind-powered generator.

Any items of switchgear and controlgear are required to be provided with a label or some other suitable means of identification, to indicate their purpose, except where the purpose of the switchgear or controlgear is obvious (Regulation 514.1.1 refers).

7.2 Identification of conductors

Regulation 514.6 requires that where identification is not required, cores of cables shall be identified by one of the following methods:

- by colour as required by Regulation 514.4 and/or
- using lettering and/or numbering as required by Regulation 514.5

7.2.1 Single-phase installations

The line conductors of single-phase circuits should be identified by the colour brown. A neutral conductor, where identified by colour, must be blue (Regulation 514.4.1 refers). Protective conductors must be identified by the green-and-yellow colour combination specified in Regulation 514.4.2.

7.2.2 Identification of switch-wires

Where a two-core and earth cable with a brown and a blue conductor is used for the wiring to a lighting switch or a lighting pendant for example, both the brown and blue conductors are used as line conductors, and so the blue conductor must be suitably identified at its terminations.

Such identification will normally be accomplished by using brown sleeving at both ends of the conductor. Alternatively, a cable with both cores in brown may be used for such a purpose.

Where a flat three-core and earth cable with cores coloured brown, black and grey is used to connect two-way or two-way and intermediate lighting switches, all three conductors are line conductors, and so the black and grey conductors must be marked appropriately (see Fig 7.1).

7.2.3 Extensions, alterations or repairs to an existing single-phase installation

Regulation 514.1.3 states that where an addition or an alteration is made to an existing single-phase installation wired in the old colour coding system, marking at the interface, such as at the double socket-outlet shown in Fig 7.2 for example, need not be necessary provided that:

- The existing cables are correctly identified by the colours of red for line conductor and black for neutral conductor.
- The new cables are correctly identified by the colours of brown for line conductor and blue for neutral conductor.

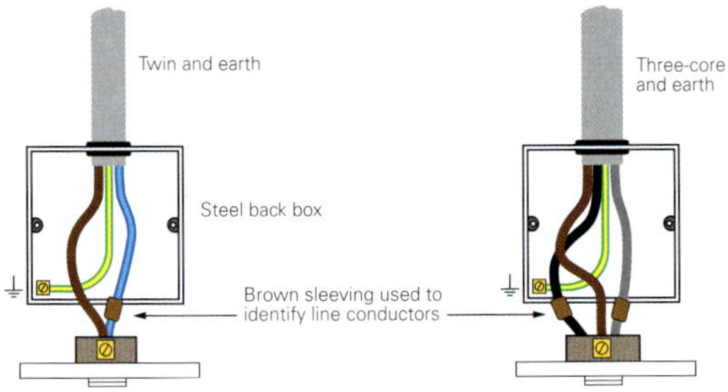

Fig 7.1 Identification of switch-wire conductors

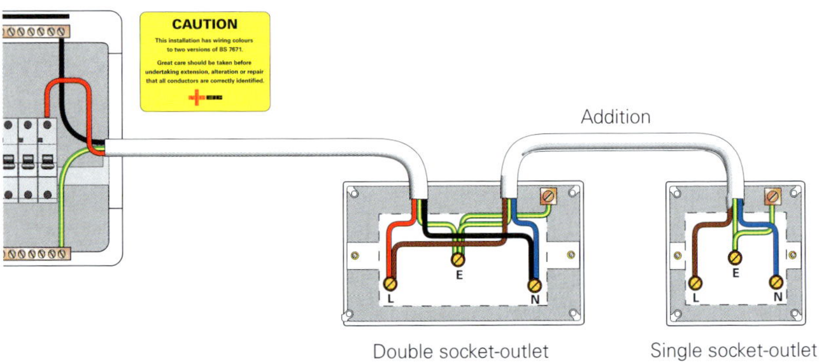

Fig 7.2 Requirements for a single-phase installation

7.2.4 Identification of conductors - warning notice at the consumer unit or distribution board

If an extension, alteration or repair is made to an installation such that both brown and blue, and red and black, cable cores are present, a warning notice must be affixed at or near the appropriate distribution board or consumer unit as shown in Fig 7.3 (Regulation 514.14.1 refers).

Fig 7.3 Identification warning notice of conductor colours

7.2.5 Three-phase installations

The circuits of a new three-phase installation should be wired in the colours of brown, black and grey for the line conductors, blue for the neutral conductor and the protective conductors remain green-and-yellow (see Fig 7.4). For single-phase circuits, the line conductors should be brown, irrespective of the supply phase to which they are connected.

An extension, alteration or repair to an existing installation should be wired in the colours of brown, black, grey and blue, and green-and-yellow, as applicable (see Fig 7.4).

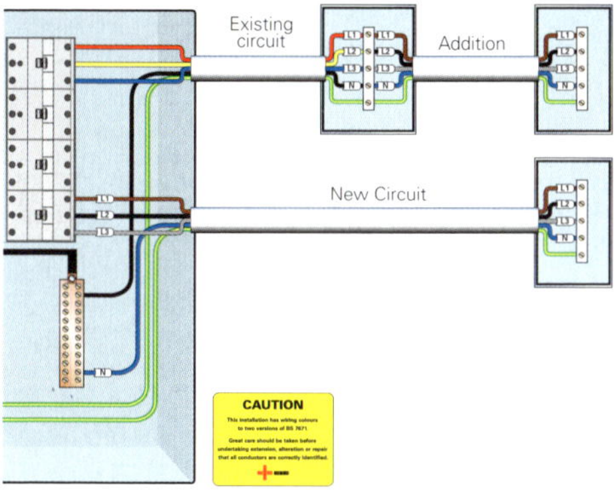

Fig 7.4 Extensions, alterations or repairs to an existing 3-phase installation

At the wiring interface(s) where an extension, alteration or repair is made with cable in the new colours to a three-phase installation wired in the old colours, unambiguous identification is required at the interface. Neutral conductors, where identified by colour, must be identified by the colour blue. Old and new line conductors should be fitted with sleeves marked L1, L2 and L3, and neutral conductors should be fitted with sleeves marked N to avoid any possibility of confusion (see Fig 7.5 a).

A permitted alternative arrangement is to use three single cores with insulation of the same colour, but unambiguous identification must be provided at the terminations. This could be achieved by using numbering or lettering (see Fig 7.5 b).

Fig 7.5 Alphanumeric identification on existing wiring and new conductors

7.2.6 Summary of conductor identification

Fig 7.6 summarises the permitted cable colours for AC low voltage and extra-low voltage conductors.

Identification of conductors

Function	Colour	Alpha numeric
Protective conductors	Green-and-yellow	
Functional earthing conductor	Cream	
AC power circuit[1]		
Line of single-phase circuit	Brown	L
Neutral of single- or three-phase circuit	Blue	N
Line 1 of three-phase circuit	Brown	L1
Line 2 of three-phase circuit	Black	L2
Line 3 of three-phase circuit	Grey	L3
Control circuits, ELV and other applications		
Line conductor	Grey	L
Brown / Orange	White	
Black / Yellow	Pink, or	
Red / Violet	Turquoise	
Neutral or mid-wire[2]	Blue	N or M

NOTES: [1] Power circuits include lighting circuits.
[2] An earthed PELV conductor is blue.

Fig 7.6 Summary of permitted means of conductor identification

7.3 Protective devices

Regulation 514.8.1 requires protective devices to be arranged and identified so that the circuit protected will be easily recognised. Manufacturers of consumer units and distribution boards usually provide a space for circuit labels either adjacent to each protective device or inside the external cover. Circuit identification must be written clearly and should provide sufficient information to ensure that it will not be confused with any other circuit on the distribution board. A circuit chart should also be provided.

7.4 Charts, diagrams, tables and circuit information

Regulation 514.9.1 requires a legible diagram, chart or table or equivalent form of information to be provided within or adjacent to each distribution board (including consumer units). This requirement may be met for example by using the NICEIC Circuit Details label which may be attached to the front of the consumer unit (see Fig 7.7).

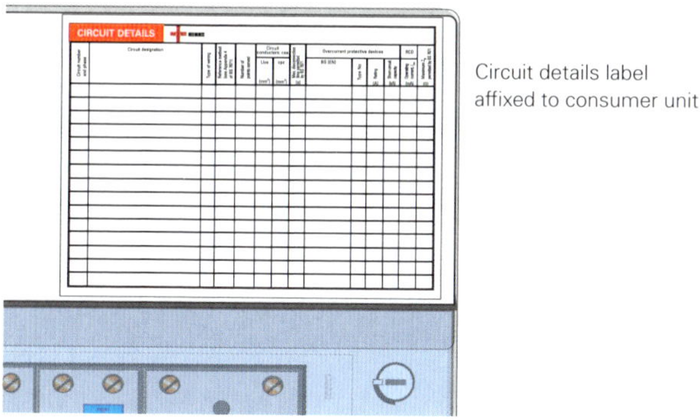

Fig 7.7 Use of NICEIC circuit details label

7.5 Voltage

Where the nominal voltage to earth (U_0) exceeds 230 V and the presence of such a voltage would not normally be expected, a warning notice is to be provided, indicating the maximum voltage present (Regulation 514.10.1 refers). The notice is to be so arranged that before access is gained to a live part, a warning of the maximum voltage to earth present is clearly visible.

7.6 Isolation

A notice of durable material is to be fixed in each position where there are live parts that are not capable of being isolated by a single device (Regulation 514.11.1 refers).

Unless there is no possibility of confusion due, for example, to the close proximity of an isolator to the equipment that it controls, the location of each isolator (or disconnector) is to be clearly identified.

7.7 Periodic inspection and testing

Regulation 514.12.1 requires a notice to be fixed in a prominent position at or near the origin of each installation stating:

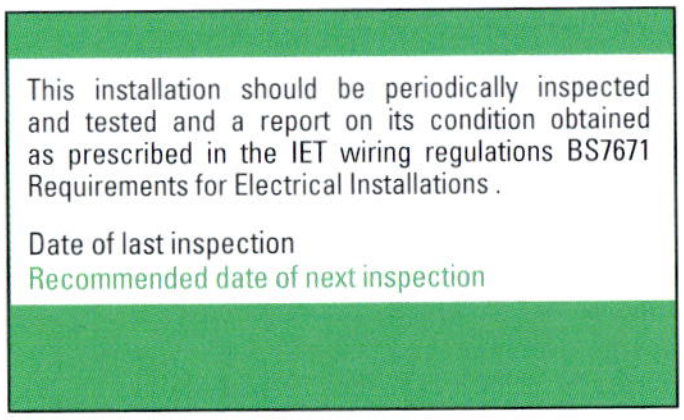

Fig 7.8 Periodic inspection and testing notice

The notice should include the date of the last inspection and the recommended date for the next inspection.

7.8 RCD test button

Regulation 514.12.2 requires a label having the wording shown below to be placed in a prominent position at, or close to, the origin of the installation.

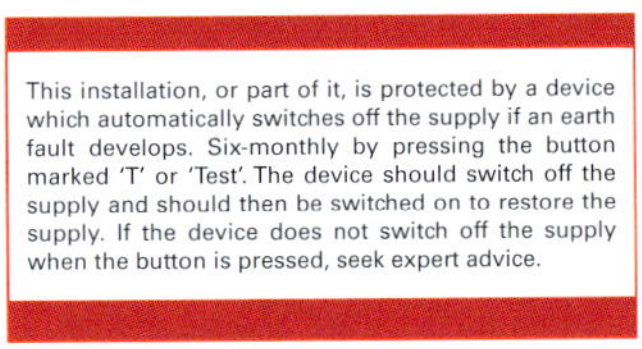

Fig 7.9 RCD test button notice

Such a notice may also be fixed in close proximity to an RCD, wherever located within the installation.

Note: Whilst not currently a requirement the presence of a notice at (or near) the origin ideally should also be confirmed for an installation which incorporates one or more arc fault detection devices installed in accordance with Regulation 532.6.

7.9 Earthing and bonding connections

In accordance with Regulation 514.13.1, a permanent label to *BS 951: 2009* having the words; "Safety Electrical Connection - Do Not Remove" (see Fig 7.10) must be permanently fixed in a visible position at or near:

- the point of connection of every earthing conductor to an earth electrode, and
- the point of connection of every bonding conductor to an extraneous-conductive-part, and
- the main earth terminal, where separate from main switchgear.

A suitable label complying with *BS 951* is normally supplied with each earthing and bonding clamp conforming to that standard. Where such a clamp is used, the label should normally be fitted to the clamp in accordance with the manufacturer's instructions.

Earthing and bonding connections made by means other than a *BS 951* clamp will require a separate label, meeting the requirements of Regulation 514.13.1, to be fixed adjacent to each connection.

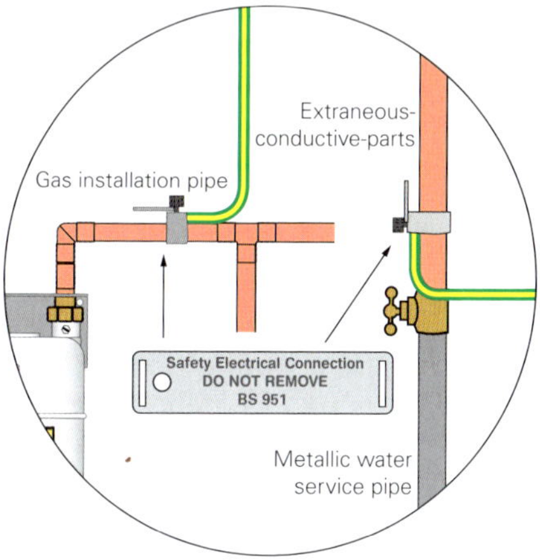

Fig 7.10 'Safety Electrical Connection - Do Not Remove' notice

Aluminium may be susceptible to electrolytic corrosion in damp conditions when in contact with dissimilar metals. Risk of electrolytic corrosion will be reduced if aluminium labels are not placed in contact with dissimilar metals.

7.10 Alternative supplies

As required by Regulation 514.15.1, where an installation includes a generating set such as a solar photovoltaic system that is used as an additional source of supply in parallel with the distributor's supply, warning notices should be affixed at the following locations in the installation:

- at the installation's origin,
- at the meter position, where this is remote from the origin,
- at the consumer unit or distribution board to which the generating set is connected,
- at all points of isolation of all sources.

The warning notice is required to have the wording given in Fig 7.11.

Fig 7.11 Warning label required where installation has alternative supplies

7.11 Circuits having a high protective conductor current

Regulation 543.7.1.205 requires information to be provided at the consumer unit to identify any circuits that have a high protective conductor current.

7.12 Presence of solar photovoltaic (PV) systems

A label should be affixed to any junction boxes forming a part of a solar PV installation warning that parts within the boxes may remain live after isolation from the PV convertor (Regulation 712.537.2.2.5.1 refers).

CHAPTER 8
DISTRIBUTION CIRCUITS

8.1 Introduction

A distribution circuit is a circuit that supplies a distribution board or other switchgear and does not extend directly to current-using equipment or a means of connection for such equipment.

An example of a distribution circuit that may be encountered in a domestic installation is a supply to another building, such as, for example, a detached garage.

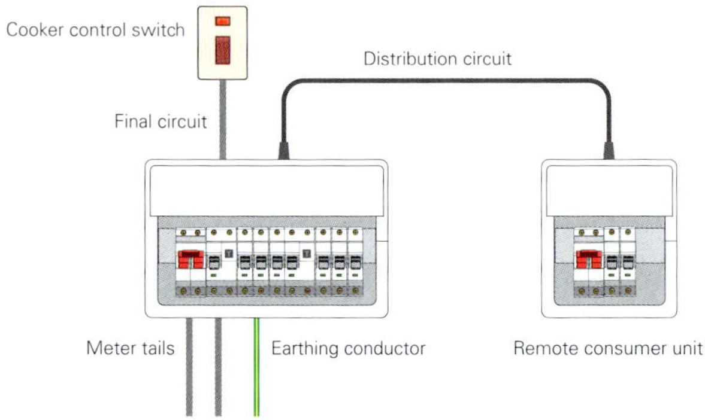

Fig 8.1 Example of a distribution circuit

8.2 Load centres and supplies to other buildings

The circumstances relating to distribution circuits in electrical installations in dwellings are diverse, and it is not possible to identify examples for all such circumstances.

However, the following two examples may serve to provide an indication where such circuits may be used to advantage.

- A distribution circuit feeding a distant internal distribution board or consumer unit from a switch-fuse or circuit-breaker at the origin of the installation. The use of such a circuit may be advantageous where the electrical loads are some considerable distance from the origin.
- A distribution circuit feeding a detached building located at some distance from the intake position in the dwelling.

It is a requirement of Regulation 411.3.1.2 that in each installation, main protective bonding conductors complying with Chapter 54 connect extraneous-conductive-parts to the main earthing terminal. This regulation also requires that where an installation serves more than one building, main protective bonding conductors must connect to the extraneous-conductive-parts in each building. A detached garage may be one example of such a building.

Where metallic pipes enter a building and have an insulating section at their point of entry, protective equipotential bonding should not be carried out (Regulation 411.3.1.2 refers).

The significance of Regulation 411.3.1.2 in the context of a building that is supplied via a distribution circuit is that the protective conductor associated with the distribution circuit supplying such another building may also serves as a main protective bonding conductor and should have a cross-sectional area complying with Regulation 544.1.1.

For an installation where PME conditions do not apply (TN-S or TT system), the copper equivalent cross-sectional area of the main protective bonding conductors must be not less than half that required for the earthing conductor of the installation, subject to a minimum of 6 mm^2. The cross-sectional area need not exceed 25 mm^2 where the bonding conductor is of copper.

Where PME conditions apply (TN-C-S system) if:

- metallic gas, water or other pipes extend to other buildings, or
- the building supplied by a distribution circuit contains extraneous-conductive-parts (such as metallic structure in contact with Earth).

Except where the next paragraph applies, the associated protective conductor of the distribution circuit must have the cross-sectional area not less than that required by Table 54.8 of *BS 7671* in relation to the PEN conductor of the distributor's supply cable. For supply neutral conductors with a copper equivalent cross-sectional area of not more than 35 mm^2, the minimum cross-sectional area of copper main protective bonding conductors need not exceed 10 mm^2.

Alternatively, the earthing system for the remote building could be segregated from the PME earthing terminal and connected to an installation earth electrode forming part of a TT earthing system, all other requirements of *BS 7671* for a TT system being met.

Reference should also be made to Chapter 2 of this, which looks at requirements for earthing and bonding in greater detail.

8.3 Assessment of current demand

When designing a new installation, the following two items will need to be considered by the installation designer:

i) Total connected load – the total electrical rating of all devices such as lamps, electrical heaters and the like that are to be connected to a distribution system.

ii) Maximum demand – the largest current normally carried by connected items of equipment, switches and protective devices when operational in normal use; it does not include the levels of current flowing under overload or fault conditions.

A consumer unit or a distribution board must be of sufficient rating to carry the total connected load. The cable supplying a final circuit must be capable of carrying the full-load current indefinitely without deterioration. However, a distribution cable may be sized to carry the maximum demand expected in normal operation on the basis that not all of the total connected load will be switched on simultaneously.

Alternatively, where a diversity factor is known from experience of a particular type of installation, the current demand of a distribution circuit may be calculated by applying a diversity factor to the sum of the connected loads.

Other methods of calculating the current demand of a distribution circuit may be used. Such methods are often based on known links between current demand and other criteria, such as floor area and population.

8.4 Voltage drop

Having determined the design current for a distribution circuit, a suitable conductor size must be chosen. One factor to be considered when choosing a suitable conductor size is the maximum value of voltage drop that can be permitted. Within *BS 7671* the requirements for voltage drop in a consumers' installation is given in Section 525, and guidance on complying with these requirements is given in Section 6.4 of Appendix 4. It should be borne in mind that for an installation that includes a distribution circuit supplying a final circuit, the total voltage drop will be the sum of the voltage drops of the distribution circuit and the final circuit.

8.5 Disconnection times

For a distribution circuit, the following disconnection times are permitted for AC circuits supplied at a nominal voltage (U_0) of 230 V.

Table 8.1 Disconnection times for distribution circuits

System earthing type	Maximum permitted disconnection time	Relevant Regulation in *BS 7671*
TT	1 second	411.3.2.4
TN	5 seconds	411.3.2.3

8.6 Recording of circuit details

As with all circuits, distribution circuits are required to be inspected and tested and the results recorded on the Electrical Installation Certificate (see Chapter 14).

Regulation 514.9.1 and Section 644 call for circuit details and test results to be provided for all circuits in an electrical installation. However, there is sometimes a tendency to overlook the provision of this information for distribution circuits (or sub-mains as they are sometimes called), where such circuits form part of the installation. The requirement to provide such information applies to distribution circuits as much as for final circuits.

One situation where this type of omission of data occurs is where a circuit-breaker, located at the origin, feeds a single distribution circuit to a remote distribution board or consumer unit (see Figure 8.1).

Without this recorded data the certification for the installation is incomplete. This does not apply to meter tails making the short connection between the consumer unit or main switch to the electricity meter or distributor's isolator.

CHAPTER 9
FINAL CIRCUITS

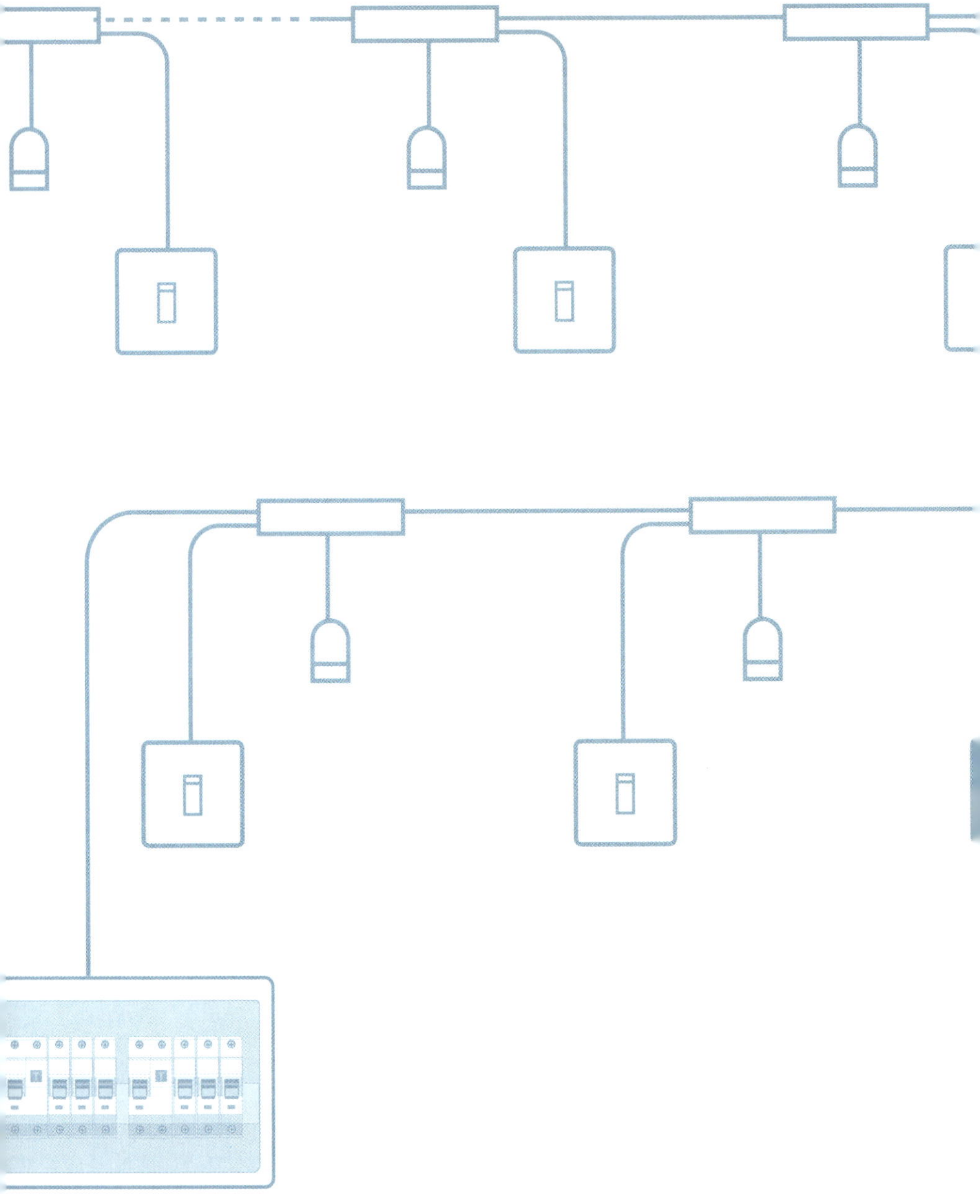

9.1 Introduction

A final circuit is a circuit that connects directly to items of current-using equipment, or to a socket-outlet or connection unit or the like for the connection of current-using equipment. Information is given in this Chapter on types of conventional circuit arrangements for use in domestic premises.

9.2 Circuits supplying socket-outlets

Socket-outlets may be supplied by either a ring or a radial final circuit. In either case, a number of items of permanently connected equipment may also be supplied by the circuit. Such equipment should preferably be of a rating not exceeding 2 kW.

Permanently connected equipment may be connected directly to a ring final circuit using a fused connection unit incorporating a *BS 1362* fuse (not exceeding 13 A) or a type B or C circuit-breaker (not exceeding 16 A) complying with *BS EN 60898* or *BS EN 61009-1*, or *BS EN 60947-2*. In the case of some existing installations, type 1, 2 or 3 MCBs complying with *BS 3871* may still be encountered (see also Chapter 3 of this Guide).

A switch should be provided where a fixed or stationary appliance that may give rise to a hazard in normal use is connected to the supply other than by a plug and socket-outlet. (Regulation 463.1.3 refers)

Socket-outlets and fused connection units may be connected to the circuit directly or via a spur (see Figures 9.1 and 9.2).

Except where the permitted exception exists, Regulation 411.3.3 requires that additional protection by means of an RCD with a rated residual operating current ($I_{\Delta n}$) not exceeding 30 mA must be provided for socket-outlets with a rated current not exceeding 32 A.

The number and positions of socket-outlets should be such as to ensure compliance with Regulation 553.1.7. This regulation stipulates that where mobile equipment is likely to be used, sufficient provision, in terms of number of socket-outlets and indeed the location of such, should be made to allow mobile equipment to be fed from an adjacent and conveniently accessible socket-outlet. Due account should be taken of the length of flexible cable normally fitted to portable appliances and luminaires when considering such provision.

9.2.1 Radial final circuit

Whilst the term radial final circuit is not defined in Part 2 of *BS 7671* it is generally considered to be a final circuit connected to the supply at one end only (see Fig 9.1).

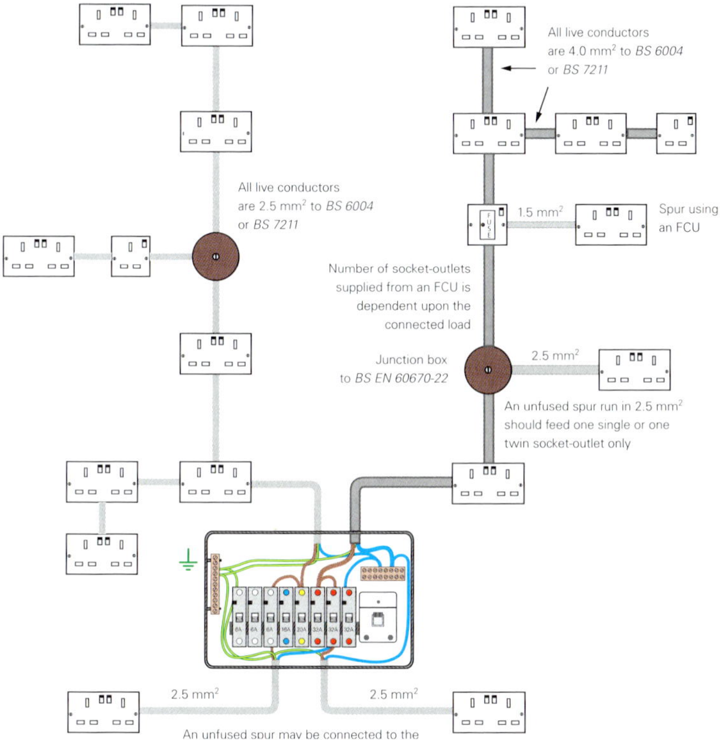

Fig 9.1 An example of a radial final circuit supplying 13 A accessories

Radial final circuits may be used to supply lighting, socket-outlets or individual items of current-using equipment such as cookers and immersion heaters.

9.2.2 Ring final circuit

A ring final circuit starts and ends at a single 'way' within a consumer unit and so two conductors will be terminated at the protective device and at the corresponding terminals on the neutral and earth bars of the consumer unit, except those connected to spurs, at each outlet (see Fig 9.2).

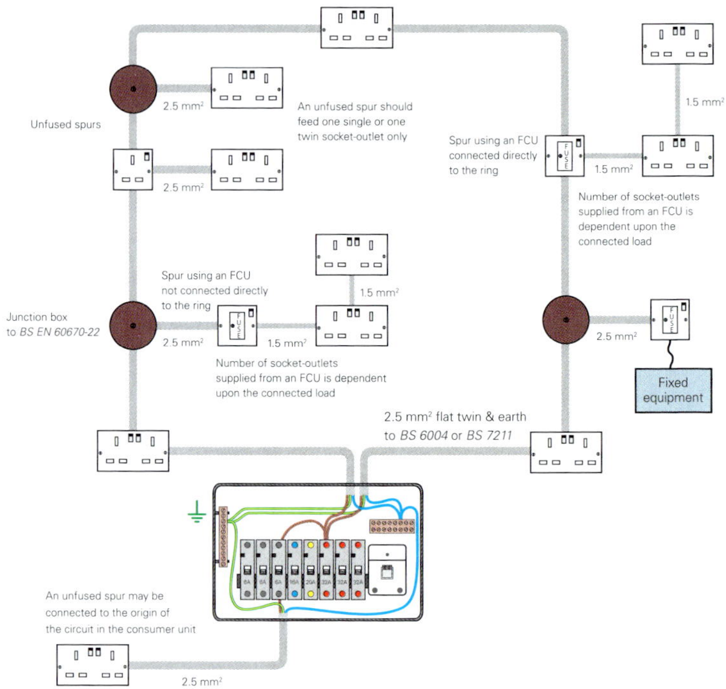

Fig 9.2 An example of a ring final circuit socket-outlets and fused connection units to *BS 1363*

In the case of ring final circuits, copper conductors should have a minimum cross-sectional area of 2.5 mm^2. For conductors of two-core mineral insulated cables, a minimum csa of 1.5 mm^2 applies. Additionally, in either case, the current carrying capacity (I_z) of the live conductors should not be less than 20 A (Regulation 433.1.204 refers).

9.2.3 Spurs

A spur is a branch from a final circuit (Part 2 of *BS 7671* refers). A spur may be either unfused or fused.

For radial final circuits, the current-carrying capacity of conductors, corrected for the particular installation conditions (I_z) used for an unfused spur, must be least equal to the rated current (I_n) of the protective device of the circuit.

For a ring final circuit each unfused spur may only feed one single or one double socket-outlet, or a single item of current-using equipment.

A fused spur originates at a fused connection unit to *BS 1363-4*, which may be of the switched or unswitched type. Conductors used for fused spurs should have an adequate current-carrying capacity for the connected load. The fused spur is protected by a cartridge fuse to *BS 1362*.

The number of socket-outlets supplied by a fused spur is unlimited. However, an assessment should be made to ensure that the current demand of the equipment, connected via the socket-outlets forming part of the fused spur, does not exceed the rating of the fuse in the connection unit in normal use.

9.3 Permanently connected loads

A radial circuit dedicated to a particular load is generally provided where the operating current of the load is such that it would dominate a circuit. For example, in most cases, a fixed electric heater rated at about 2 kW or more is served by a dedicated circuit.

Where a number of items of current-using equipment are served by a common circuit, disconnection of all the items will occur if the circuit protective device operates.

Where disconnection of an item of equipment under such circumstances is likely to cause inconvenience or danger, a dedicated circuit should be provided for that equipment so that disconnection occurs only in the event of a fault attributable to that equipment or its associated circuit.

9.4 Introduction to lighting circuits

In this Guide, in order to avoid fragmentation, the requirements of Section 714 (Outdoor lighting installations) are included in the lighting section of Chapter 12 (Electrical equipment outside the dwelling) and the requirements of Section 715 (Extra-low voltage lighting installations) are included in Section 9.4.3 of this Chapter. A lighting circuit may also contain smoke and / or heat alarms as part of a Grade D fire alarm system according to *BS 5839-6'*.

Lighting in household premises should be served by one or more designated radial final circuits, according to the size and layout of the property. It could be argued that, where practicable, more than one lighting circuit should be provided, even in small premises, so as to avoid the loss of all lighting should one circuit be disconnected due to the operation of its circuit protective device.

Similarly, it might be desirable for areas such as a hallway and stairs to be served by separate circuits so that some lighting for the stairs is still available should one of the circuits be disconnected.

Regulation 559.5.1 lists acceptable means of final connection that should be provided at each lighting point. In domestic premises this final connection is likely to be made via a ceiling rose (pendant set), a batten lampholder, or by direct connection to the terminals of a luminaire.
Each radial final circuit for lighting connects a number of such lighting points to the circuit protective device within a consumer unit or a distribution board (see Fig 9.3).

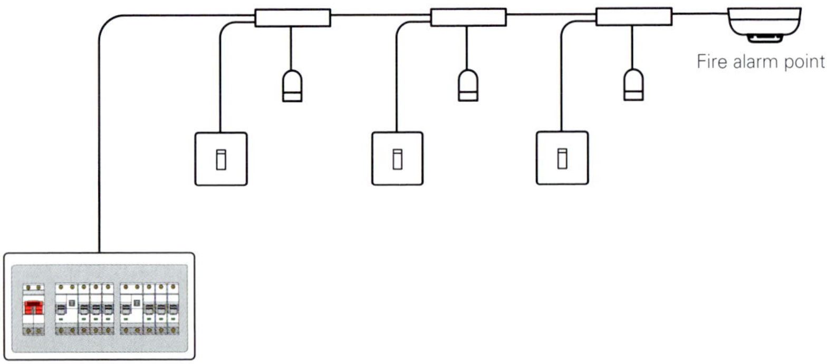

Fig 9.3 A typical radial final circuit supplying lighting

9.4.1 Ceiling roses

Where a luminaire is installed suspended by a pendant set, the method of fixing the ceiling rose to the fabric of the building should be capable of carrying a mass of not less than 5 kg, and be in accordance with the manufacturer's instructions. The weight of the luminaire and its accessories must be compatible with the mechanical capability of the ceiling or suspended ceiling or supporting structure where installed (Regulation 559.5.2 refers).

Where ceiling roses are used, Regulation 559.5.1.202 requires that only one outgoing flexible cable be connected to each ceiling rose, except where the ceiling rose is specifically designed for more than one such pendant connection.

It should be noted that Regulation 559.5.1.201 precludes the use of ceiling roses in circuits that normally exceed 250 V in operation. Ceiling roses complying with *BS 67: 1987* have maximum current and voltage ratings of 6 A and 250 V respectively, and are intended for use in final circuits having a rating not exceeding 16 A where the ceiling rose has screw-type supply terminals, or not exceeding 10 A where the supply terminals of the ceiling rose are the screwless type.

9.4.2 Luminaire supporting couplers

Where there is a requirement for a luminaire to be easily disconnected from the fixed wiring installation, for maintenance or cleaning purposes for example, this may be achieved by using a luminaire supporting coupler (LSC) (see Fig 9.4) as the means of connecting the luminaire to the fixed wiring system.

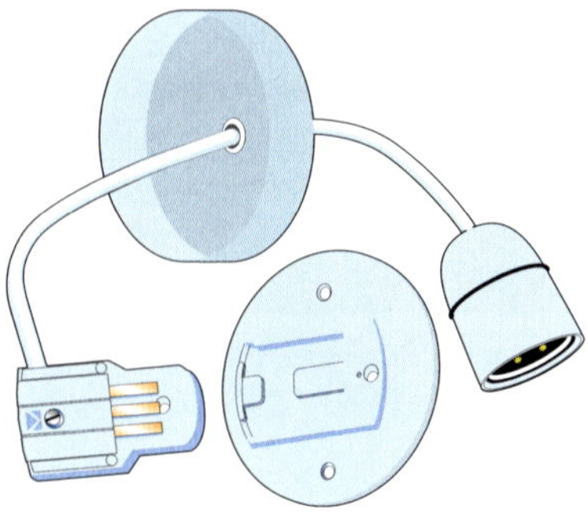

Fig 9.4 Typical luminaire supporting coupler (LSC)

Each LSC comprises an LSC outlet for fixed mounting and connection to the fixed wiring installation, and an LSC plug which is either connected to the flexible cable of a luminaire, or which is a component part, or an integral part, of the luminaire.

Not all LSCs are interchangeable; that is, the plug of one LSC may not be compatible with the outlet of another. Where there is a requirement for interchangeability so that, for example, a complete luminaire may be quickly unplugged and replaced or moved to another position, an LSC complying with *BS 7001* should be used.

LSCs complying with *BS 7001* are of a standardised format and are interchangeable, irrespective of the manufacturer, whereas those complying with *BS 6972* are not necessarily so.

Ceiling roses to *BS 67*, commonly used in household premises, and luminaire supporting couplers to *BS 6972* or *BS 7001* have a rated current not exceeding 6 A, and lighting circuits for household premises are generally protected by a protective device with a rated current or current setting (I_n) of 6 A.

Historically, for the purposes of designing the lighting circuit(s) of domestic premises, each lighting point was considered to have a rating of 100 W, which meant a typical connected load of 10 single lighting points to a circuit. With the advances made in lighting technologies and the focus centred on energy efficiency, manufacturers have produced a range of 230 V lamps having ratings of between 3 W and 50 W. These give a similar light output as was previously available from 40 W, 60 W and 100 W incandescent lamps. For example, a typical 10 W LED BC lamp currently has an equivalent light output comparable to a 60 W incandescent lamp. This permits a greater flexibility in lighting design since the number of lighting points that can be connected to say, a 6 A circuit, does not have the same restriction there once was.

Where luminaires incorporating discharge lamps (such as compact fluorescent) are used, account should be taken of the load of their control gear and the transient and harmonic currents produced when assessing the load of the lighting circuit.

The design of lighting circuits must incorporate the means to switch individual luminaires or groups of luminaires on and off according to their location (Regulations 537.3.1.1, 537.3.1.2 and 537.3.1.3 refer). The switching arrangement (whether one-way, two-way or intermediate) should be so placed to be readily accessible at the points of access to the area being illuminated.

With reference to Table 52.3 of *BS 7671*, the minimum conductor size for a radial final circuit for lighting, protected by an overcurrent protective device with a rated current or current setting (I_n) of 6 A, is 1.0 mm^2 for thermoplastic (PVC) or thermosetting insulated cables with copper conductors.

The flexible cable between the ceiling rose or similar and the lampholder is permitted to have a minimum cross-sectional area of 0.75 mm^2 (see Regulations 433.3.1 (ii), 524.1 and Table 52.3). This flexible cable should preferably have 90 °C thermoplastic insulation.

As for any other circuit, the voltage drop between the origin of the installation and the terminals of the luminaire needs to be checked in order to verify compliance with Regulation Group 525 and Appendix 4 of *BS 7671*.

For 70 °C thermoplastic (pvc) insulated and sheathed flat cable with protective conductor, examples of the maximum circuit cable length, consideration having been given to voltage drop (amongst other things), are given in Appendix A of this Guide.

9.4.3 Extra-low voltage lighting installations

Extra-low voltage (ELV) lighting installations, which typically are used to supply extra-low voltage LED or tungsten halogen lighting, are subject to the requirements of Section 715. When dealing with extra-low voltage lighting installations, it should be remembered that the content of Section 715 modifies, or is additional to, the content of Parts 1 to 6 of *BS 7671* which means that, where relevant, the content of Section 559 still applies to the extra-low voltage lighting installations.

Section 715 applies to lighting installations supplied from a source with a maximum rated voltage of 50 V AC or 120 V DC (Regulation 715.1 refers).

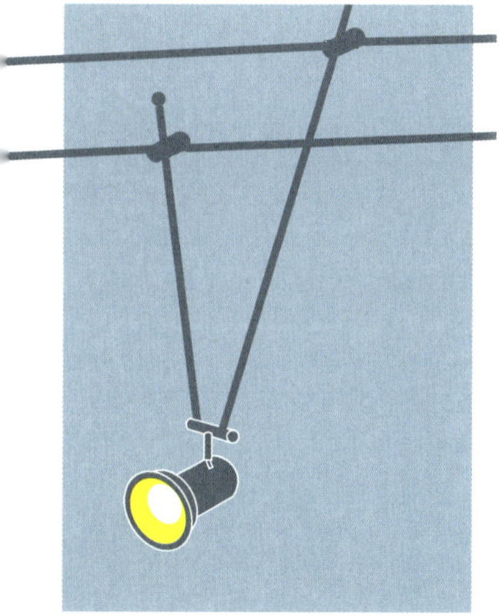

Fig 9.5 Suspended extra-low voltage

For safety reasons, whilst Section 715 permits the extra-low voltage system used to supply the lighting installation to be either separated extra-low voltage (SELV) or protective extra-low voltage (PELV), functional extra-low voltage (FELV) is not to be used (Regulation 715.411.7.201 refers).

Where extra-low voltage luminaires are installed without provision for the connection of a protective conductor, they should only be installed as part of a SELV system. (Regulation 715.414 refers).

As shown by Fig 9.5, some extra-low voltage systems employ, on the output of the transformer, uninsulated conductors and, for this arrangement, the maximum operating voltage is reduced to 25 V AC or 60 V DC (Regulation 715.414 refers).

In ELV lighting systems the LED lighting systems and some tungsten halogen systems operate using an extra-low voltage supply which is produced by either a transformer or electronic convertor conforming to the appropriate British Standard specified in Regulation 715.414.

The transformer or convertor is normally supplied by a standard 230 V mains supply (see Fig 9.6).

Transformers can only provide an AC supply, whereas electronic convertors depending upon their design, can provide either an AC or DC supply. An electronic convertor has the additional advantage that when there is a change in either the current taken by the lamp(s) or in the input voltage, the electronic convertor will try to keep the voltage to the lamps at the rated value.

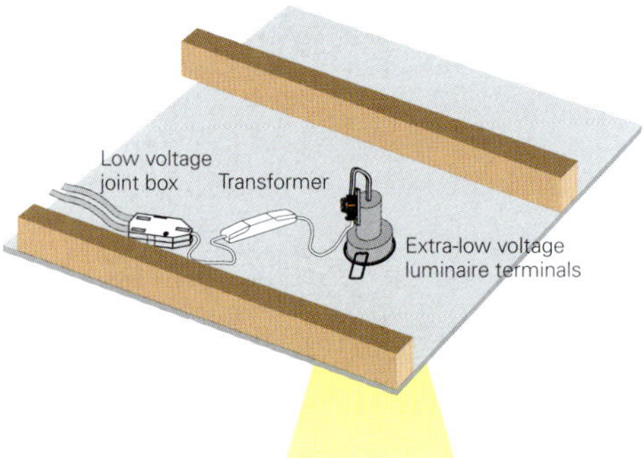

Fig 9.6 Typical installation of an extra-low voltage luminaire in a suspended ceiling

As listed in Table 55.2 of *BS 7671* the symbols for a short-circuit proof safety isolating transformer and an electronic convertor, for use in an ELV lighting installation are shown in Fig 9.7.

a) Symbol for short-circuit proof safety isolating transformer

b) Symbol for electronic convertor for in an ELV lighting installation

Fig 9.7 Symbols for short-circuit proof safety isolating transformer electronic convertor for use in ELV lighting installations

If a short-circuit occurs across the output of a transformer, there is the possibility that the heat generated by the short-circuit current can be the cause of a fire. To protect against such an event Regulation 715.422.106 requires the transformer / convertor used to supply an extra-low voltage lighting system should either employ a fail-safe overcurrent protective device that meets all the features described in Regulation 715.422.107.2 or be short-circuit proof (be able to withstand the effects of a short circuit).

Where, to protect against the effects of short circuit, the transformer employs an overcurrent protective device, Regulation 715.430.104 limits the use of self-resetting overcurrent protective devices to transformers up to 50 VA. For overcurrent protective devices protecting transformers above this rating, once the device has operated, it must be manually reset.

The likelihood of a short-circuit occurring is particularly high for extra-low voltage systems that use un-insulated conductors on the secondary side of the transformer. Where both live conductors are un-insulated, Regulation 715.422.107.1 requires that, either:

- they should be provided with a protective device complying with the requirements of Regulation 715.422.107.2, or
- the system should comply with *BS EN 60598-2-23*, or
- the maximum rating of the transformer or convertor should not exceed 200 VA.

Other regulations relating to protection against thermal effects and the impact upon the fire performance of the building can be found in Regulations 421.1.2, 421.1.4 and Section 527.

Regulation 715.521.1 permits a wide range of wiring systems to be used for ELV lighting systems. Track systems and systems where the lamp is suspended from bare conductors allow the position of the lamps to be altered.

Where a system using bare conductors is installed it is not permissible to utilise metallic structural parts of the building or parts of furniture as live conductors. Where parts of the ELV lighting installations are accessible, the maximum temperature limits given in Table 42.1 of *BS 7671* should not be exceeded.

Where un-insulated live conductors are used the following requirements apply:

i) the nominal voltage does not exceed 25 V AC or 60 V DC

ii) the cross-sectional area of the conductor is not less than 4 mm^2 (see Table 9.1), and

iii) the conductors are not placed directly on combustible material.

In addition, where bare conductors are used, in order to prevent a short-circuit in the section between the transformer and the short-circuit protection device, at least one conductor and its terminal should be insulated.

Regulation 715.521.107 contains requirements for the use of suspended ELV lighting systems.

Included in this Regulation are the following requirements:

i) where the ELV system is suspended, the suspension devices must be able to support five times the weight of the ELV luminaire and its lamp and not less than 5 kg,

ii) where the suspended system is fixed to the walls or ceilings by distance cleats the installation should be accessible throughout its length,

iii) the use of insulation piercing type connectors is prohibited.

Operating at extra-low voltage means that the cable between the transformer or convertor and the lamps can carry a comparatively high current. Irrespective of the lamp rating or the type of lamp Regulation 715.524.201 specifies the minimum cross-sectional area of the live conductors.

In the case of systems with luminaires suspended from the conductors, the minimum cross-sectional area of the ELV conductors which are connected to the output terminals or terminations of transformers / convertors shall be 4 mm^2 for mechanical reasons.

For extra-low voltage lighting installations, in order to comply with Regulation Group 525, the voltage drop between the transformer / electronic convertor and furthest luminaire should not exceed 5 % of the nominal voltage of the ELV installation; that is 0.6 V for a 12 V installation (Regulation 715.525 refers).

Regulation 715.530.3.104 requires that for ELV lighting systems the protective devices are accessible.

For installations where the luminaires of ELV lighting systems are recessed into suspended ceilings and the control gear and wiring is concealed above the ceiling (as shown in Fig 9.6), following the operation of the protective device accessing the protective device can be problematic. Regulation 715.530.3.104 requires that where the control gear for ELV lighting is located above a false ceiling, information regarding the presence and location of the protective devices is provided.

Regulation 715.530.3.104 also requires that:

i) the SELV sources, protective devices and similar equipment shall be permanently connected, adequately supported and mounted in such a way that mechanical stress on the terminations is avoided, and

ii) the SELV sources and their protective devices shall be installed in such a way that overheating of the SELV sources and their protective devices due to thermal insulation is avoided.

To enable effective isolation where an ELV lighting system is fed from several transformers connected in parallel, the primary circuits shall be permanently connected to a common isolating device (Regulation 715.537.1.1 refers).

9.5 Voltage drop

Where a circuit is supplied from a distribution board that is not at the origin of the installation, the total voltage drop between the origin of the installation and the load end of the circuit is equal to the sum of the voltage drops in the distribution circuit and the circuit concerned (normally a final circuit). For a single-phase installation in domestic premises, this is a straightforward arithmetic sum and requires no further explanation.

The requirements of Regulation Group 525 of *BS 7671* are deemed to be satisfied if (as given in Section 6.4 of Appendix 4 in *BS 7671*) the voltage drop between the origin of the installation and either the fixed current using equipment or the terminals of a socket-outlet does not exceed 5 % of the nominal voltage and in the case of lighting circuit does not exceed 3 % of the nominal voltage.

CHAPTER 10

FACTORS RELATING TO INSTALLATION OF WIRING SYSTEMS AND CABLES

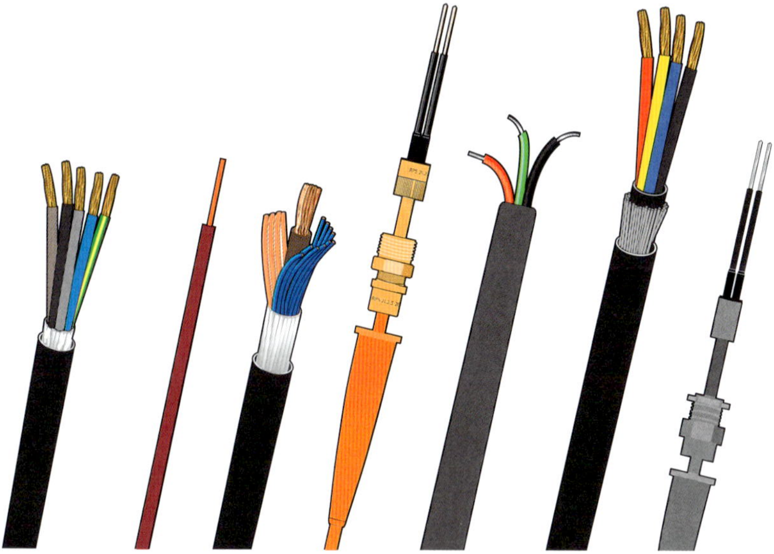

10.1 Wiring systems – types

To comply with the requirements of Regulations 511.1 and 511.2, wiring systems must utilise cables complying with the relevant requirements of an applicable British Standard or Harmonised Standard. Where equipment is used which does not comply with a relevant British Standard or Harmonised Standard or it is used outside the scope of its standard, the designer of the electrical installation must verify that the equipment provides the same degree of safety as that afforded by compliance with the Regulations.

Any prefabricated wiring systems employed should comply with *BS 8488: 2009 (2010) – Prefabricated wiring systems intended for permanent connection in fixed installations – Specification.*

Regulation 512.1.1 requires that every item of equipment (including cables) to be suitable for the nominal voltage (U_0) of the installation and cable manufacturers always state the voltage grade for particular types of cables. Conduit wiring cables (such as 6491X) are 450 / 750 V grade. Flat twin and earth wiring cables (such as 6242Y) are designated 300 / 500 V grade.

10.2 Applications of cables for fixed wiring

Table 10.1 provides information about a range of cables in common use by electrical installation contractors.

Table 10.1 The range of cables in common use

Type of cable	Uses	Comments
Thermoplastic (PVC), thermosetting or rubber insulated non-sheathed	In conduits and cable trunking	Intermediate support may be required on long vertical runs 70 °C maximum conductor temperature for normal wiring grades – including thermosetting types (Note 4). Cables run in PVC conduit are not permitted to operate with a conductor temperature greater than 70 °C (Note 4)
Flat thermoplastic (PVC) or thermosetting insulated and sheathed	General indoor use in dry or damp locations. May be embedded in plaster On exterior surface walls, boundary walls and the like. Overhead wiring between buildings (Note 6). Underground in conduits and pipes In building voids or ducts formed of building fabric	Additional mechanical protection may be necessary where exposed to mechanical stresses. Protection from direct sunlight may be necessary. See Note 4. Unsuitable for embedding directly in concrete May need to be hard drawn (HD) copper conductors for overhead wiring (Note 6)
Split-concentric thermoplastic (PVC) insulated and sheathed	General	Additional mechanical protection may be necessary where exposed to mechanical stresses. Protection from direct sunlight may be necessary. Black sheath colour is better for cables in sunlight
Mineral insulated	General	With overall PVC covering where exposed to the weather or risk of corrosion, or where installed underground, or in concrete ducts
Thermoplastic (PVC) or XLPE insulated, armoured, thermoplastic sheathed	General	Additional mechanical protection may be necessary where exposed to mechanical stresses. Protection from direct sunlight may be necessary. Black sheath colour is better for cables in sunlight
Screened multicore cables having low emission of smoke and corrosive gasses when affected by fire, to *BS 8436*	General use for use in walls, partitions and building voids	Care should be exercised during installation to avoid any damage to cable coverings

1) The use of cable covers is desirable for all underground cables which might otherwise be disturbed. Route marker tape should also be installed, buried just below ground level.
2) Cables having thermoplastic (PVC) insulation or sheath should preferably be installed only when the ambient temperature is above 0 °C, and has been for the preceding 24 hours. Where thermoplastic (PVC) cables are to be installed during a period of low temperature, precautions should be taken to avoid risk of mechanical damage during handling. A minimum ambient temperature of 5 °C is advised in *BS EN 50565* for some types of thermoplastic (PVC) insulated and sheathed cables.
3) Cables must be suitable for the maximum ambient air temperature, and must be protected from any excess heat produced by other equipment, including other cables.
4) Thermosetting cable types (to *BS 7211* or *BS 5467*) can operate with a conductor temperature of 90 °C. This must be limited to 70 °C where cables are drawn into a conduit made of thermoplastic material or where installed in a common enclosure with cables with a maximum conductor operating temperature of 70 °C. Also equipment must not be connected to conductors intended to operate at temperatures higher than 70 °C unless the manufacturer has confirmed that the equipment is suitable for such conditions. Regulation 512.1.5 refers.
5) For cables to *BS 5467*, *BS 6004*, *BS 6724* and *BS 7211* further guidance may be obtained from those standards. Additional advice is given in *BS EN 50565-1: 2014* Electric cables. to use for cables with a rated voltage not exceeding 450 / 750 V (U_0 / U). General guidance.
6) Cables used for overhead wiring between buildings must be able to support their own weight and any imposed wind loading and any ice or snow loading. A catenary support is normally used.

10.3 Installation methods

The basis for determining the current-carrying capacity of cables is set out in Appendix 4 of *BS 7671*. The value of tabulated current-carrying capacity for a particular cable is dependent upon the chosen Installation method, and this is reflected in the data of current-carrying capacities.

Table 4A2 lists the Installation methods for which consideration has been given to current-carrying capacities in Appendix 4 of *BS 7671*. Reference methods are installation methods where a determination (by test or calculation) has been made of current-carrying capacity in more specific circumstances.Table 10.2 identifies some common installation and reference methods.

Table 10.2 Common installation and reference methods for thermoplastic cables used in dwellings

Installation description	Installation method	Reference method
Non-sheathed cables in conduit in a thermally insulated wall having an inner skin of thermal conductance not less than 10 W / m^{2}K	1	A
Multicore cable in conduit in a thermally insulated wall having an inner skin of thermal conductance not less than 10 W / m^{2}K	2	A
Multicore cable direct in a thermally insulated wall having an inner skin of thermal conductance not less than 10 W / m^{2}K	3	A
Multicore cable in conduit on a wooden or masonry wall, or spaced less than 0.3 times the conduit diameter from it	5	B
Single-core or multicore cables fixed on (clipped direct), or spaced less than 0.3 times the conduit diameter from a wooden or masonry wall	20	C
Non-sheathed (50) or sheathed (51) cables in flush cable trunking in the floor	50,51	B
Non-sheathed (52) or sheathed (53) cables in flush cable trunking	52,53	B
Single-core or multicore cables direct in masonry having a thermal resistivity not greater than 2 K.m / W without added mechanical protection	57	C
Single-core or multicore cables direct in masonry having a thermal resistivity not greater than 2 K.m / W with added mechanical protection such as capping	58	C
Non-sheathed (59) cables or multicore (60) cables in conduit in masonry of thermal resistivity not greater than 2 K.m / W	59	60
Flat twin and earth cable clipped direct to a wooden joist, or touching the plasterboard ceiling surface, above a plasterboard ceiling with thermal insulation of thickness **not** exceeding 100 mm having a minimum U value of 0.1 W / m^{2}K	100	Table 4D5
Flat twin and earth cable clipped direct to a wooden joist, or touching the plasterboard ceiling surface, above a plasterboard ceiling with thermal insulation of thickness **exceeding** 100 mm having a minimum U value of 0.1 W / m^{2}K	101	Table 4D5
Flat twin and earth cable in a stud wall with thermal insulation having a minimum U value of 0.1 W / m^{2}K with the cable touching the inner wall surface, or touching the plasterboard ceiling surface , and the inner skin having a minimum U value of 10 W / m^{2}K	102	Table 4D5
Flat twin and earth cable in a stud wall with thermal insulation having a minimum U value of 0.1 W / m^{2}K with the cable not touching the inner wall surface	103	Table 4D5

Where a cable is installed in contact with thermal insulation, its current-carrying capacity might be reduced by as much as 50 % of its tabulated value, depending on the particular circumstances.

The data in Table 4D5 of *BS 7671* indicates that the current carrying-capacity (Reference method C – clipped direct) of 70 °C thermoplastic (PVC) insulated and sheathed cables will be reduced by approximately 25 % where the same cable is installed to Reference method A (enclosed in conduit in an insulated wall). For domestic lighting circuits rated at 6 A, this is unlikely to cause a problem since the smallest permitted conductor size of 1 mm^2 has a current-carrying capacity above that which is needed. However, increased cable sizes may have to be selected for circuits supplying other equipment.

For a single cable completely surrounded by thermal insulation for a length of 500 mm or more, its current-carrying capacity (clipped direct) is reduced by 50 %. Where a cable is surrounded by thermal insulation for less than 500 mm, its current-carrying capacity is reduced appropriately depending on the size of cable, the length of cable in contact with thermal insulation, and the thermal properties of the insulation. Guidance is given in Table 52.2 of *BS 7671*, which is reproduced as Table 10.3.

Table 10.3 Derating factors to be applied where a cable is surrounded by thermal insulation

Length of thermal insulation (mm)	50	100	200	400
Derating factor	0.88	0.78	0.63	0.51

The above derating factors are appropriate for conductor cross-sectional areas up to 10 mm^2 in thermal insulation having a thermal conductivity (λ) greater than 0.04 $Wm^{-1}K^{-1}$.

Consider, for example, a 6 mm^2 70 °C thermoplastic (PVC) insulated and sheathed 'twin and earth' cable installed in the loft, supplying a domestic cooker. The cable has a current-carrying capacity of 47 A at an ambient temperature of 30 °C when clipped direct, not grouped with other loaded cables, and the design current is 30 A. If the cable were surrounded with thermal insulation for more than 500 mm of its length, its current-carrying capacity would be reduced to 23.5 A. This could lead to overheating and potential danger, and possible premature ageing of cable insulation and sheath.

Either the cable would have to be installed along a different route so that it is not surrounded by thermal insulation, or a cable having higher current-carrying capacity should be used.

It should be noted that subsequent introduction of thermal insulation in a loft where cables already exist can cause cables to overheat. The actual risk will depend on the type of cable, type and thickness of thermal insulation and the electrical loading of the cables.

Where the current-carrying capacity of a cable detailed in Appendix 4 has already taken account the effects of thermal insulation, a second derating factor should not be applied.

It should also be noted that material used for sound insulation (for example in inter-floor spaces) may also be thermally insulating.

10.4 Wiring in lofts

Cables installed in a roof space may cause a tripping hazard if not correctly placed and secured. Cables could also be damaged by impact, snagging or other mechanical stresses. Cables surrounded by thermal insulation could also exceed their maximum operating temperature if not correctly selected (see Section 10.3).

Cables in lofts and other accessible voids should **not** be installed over joists (supporting the ceiling below) or over the low-level parts of roof trusses, except:

- in areas of the void where the cables will not be susceptible to mechanical damage, or
- where the cables are provided with adequate protection against mechanical damage.

It should be borne in mind that the joists and the like in lofts and other accessible voids are liable to being walked on (for example by persons using the space for storage or to gain access to water tanks), and to having objects placed on or moved across them (such as stored items or loose-laid boards). It is therefore necessary for cables to be positioned or protected so that they will not be damaged by adverse influences such as impact, penetration, abrasion, crushing or snagging.

Flat 70 °C thermoplastic (PVC) insulated and sheathed cable covered by Table 4D5 of *BS 7671* installed in accessible positions should be supported by clips at spacings not exceeding the appropriate value given in Table 1 of *BS EN 50565-1: 2014*. Table 10.8 is based on some of the information within Table 1 of that standard. Refer to Section 10.8 of this Chapter for further details.

In this Chapter advice regarding the drilling of ceiling joists can be found in Section 10.5.7 (Protection of concealed cables in floors and ceilings) and Section 10.6.1 (Penetration of elements of building construction by wiring systems).

10.5 External influences

An external influence is a factor that affects the design and safe operation of an electrical installation that relates to its use or the environment in which it is located.A list of external influences relating to Chapter 32 (Classification of external influences) of *BS 7671* is given in Appendix 5 of that Standard.

Chapter 52 of *BS 7671*, which refers to (amongst other things) the codification of external influences developed for IEC Publication 60364-3, identifies the requirements for selection and erection of wiring systems in relation to external influences.

Protection against external influences is not restricted to completed work. For example, Regulation 522.6.1 requires that wiring systems shall be selected and erected so as to minimise mechanical damage; such as damage due to impact, abrasion, penetration, compression, or tension; during installation, use and maintenance.

Chapter 51 of *BS 7671* gives the requirements for selection and erection of equipment. In short, either equipment should, of itself, be suitable for the environment in which it is to be installed or, where this is not the case, and where appropriate, additional measures should be taken during installation to provide the necessary protection (Regulations 512.2.1 and 512.2.2 refer).

10.5.1 Ambient temperature

Ambient temperature is the temperature of the air or other medium surrounding the electrical equipment.

Component parts of the electrical installation should be designed to operate in the highest and the lowest ambient temperatures to which they are expected to be subjected where they are installed.

In the UK, ambient temperature in most electrical installations indoors is likely not to exceed 30 °C except in close proximity to heating equipment, where localised higher temperature should be expected.

One area in domestic installations where the ambient temperature is likely to be higher than 30 °C is an airing cupboard. Only cables having a suitably high temperature rating should be connected to an immersion heater.

The current-carrying capacities of cables given in Appendix 4 of *BS 7671* are based on an ambient temperature of 30 °C. Where the ambient temperature of an installation is known or expected to be higher or lower than 30 °C, Tables 4B1 and 4B2 give rating factors to be applied to the tabulated values of current carrying-capacity.

Thermoplastic (PVC) insulated and sheathed cables, including flat cables with protective conductors, and certain other types of cable, should be installed only when the cable and the ambient temperature is above 0 °C and has been so for the previous 24 hours.

10.5.2 External heat sources

External heat sources such as sunlight and radiant heat from certain types of equipment can raise the temperature of cables and in some cases cause degradation.

Cable sheaths containing carbon black and other types of ultraviolet (UV) stabilisers are less likely to be affected by solar radiation. However, shielding is always recommended. Where radiated heat is sufficiently powerful to raise the temperature of cables above ambient temperature, a correction factor should be applied to the tabulated current carrying-capacity. See also Section 10.5.17 of this Guide.

10.5.3 Water and high humidity

A correctly selected and erected wiring system must be installed such that no damage will be caused by condensation or ingress of water during installation, use and maintenance.

Protection against the effects of condensation can be achieved by provision of suitable drainage holes in wiring accessories at low points within the installation. Typically, the use of a hole having a diameter of 5 mm is adequate.

Only suitably designed equipment should be exposed to rain, water spray, or immersion, and such suitability should confirmed by the manufacturer's data.

10.5.4 Solid and foreign bodies

A wiring system should be selected and erected to minimise the ingress of solid foreign bodies during installation, use and maintenance. Such ingress may be prevented during installation by the use of temporary covers.

10.5.5 Corrosive and polluting substances

Parts of the wiring system and other equipment likely to be in contact with corrosive and polluting substances must be either protected or manufactured from materials resistant to such substances.

Equipment outdoors must be resistant to corrosion and be weather resistant. Metals liable to initiate electrolytic action should not be placed in contact with each other.

10.5.6 Impact

Wiring systems must be selected and erected so as to minimise mechanical damage for example damage due to impact, abrasion, penetration, compression, or tension, during installation, use and maintenance.

Additional mechanical protection should be provided to any part of the electrical installation which may be vulnerable to mechanical damage. For example, localised protection may be required to protect cables against impact from, for instance, footwear or a floor cleaning appliance.

The following sections of this Guide relate to protection against impact:

10.5.7 Protection of concealed cables in floors and ceilings.
10.5.8 Protection of concealed cables in walls.
10.5.9 Protection of cables embedded in plastered walls.

Regulation 522.6.201 relating to wiring systems located under a floor or above a ceiling and Regulation 522.6.202 and 522.6.203 relating to wiring systems concealed in walls can be satisfied by using one of the installation methods described in Regulation 522.6.204 and summarised in Table 10.4

Table 10.4 Methods of installation permitted by Regulation 522.6.204

Type of cable	Uses
(i)	Incorporate an earthed metallic covering, capable of satisfying the requirements for a protective conductor, and comply with one of the following *BS 5467, BS 6724, BS 7846, BS EN 60702-1* or *BS 8436*. (Details of the metallic-covered cables that can be used to satisfy this requirement are given in Table 10.5 of this Guide).
(ii)	Be enclosed in earthed conduit, capable of satisfying the requirements for a protective conductor, complying with *BS EN 61386-21*
(iii)	Be enclosed in earthed trunking or ducting, capable of satisfying the requirements for a protective conductor complying with *BS EN 50085-2-1*.
(iv)	Be mechanically protected in a manner sufficient to penetration of the cable by nails, screws and the like.
(v)	Form part of a SELV or PELV circuit meeting the requirements of Regulation 414.4.

Meeting the requirements of Regulation 522.6.204

As permitted by Regulation 543.2.5 options (i) to (iii) of Regulation 552.6.204 allows the use of either the earthed metallic covering of a cable, an earthed conduit or earthed trunking to be used as a protective conductor. However, the protective conductor needs to be suitable for the purpose and satisfy the requirements of indents (i) and (ii) of Regulation 543.2.2, relating to electrical continuity and cross-sectional area respectively.

Regulation 522.6.204 Option (i) cable incorporating an earthed metallic covering

The use of the metallic covering of a cable to satisfy option (i) of Regulation 522.6.204 is intended to ensure that it will not be possible for a nail, screw or the like to make contact with a line conductor of the cable without first coming into good electrical contact with the earthed metallic covering (by piercing). A direct path is thus provided for earth fault current to flow and cause automatic disconnection of the circuit concerned.

Those metallic-covered cables that can be used to satisfy the requirements of Regulation 522.6.204 (i) are listed in Table 10.5.

Table 10.5 Type of metallic covered cables that may be used to meet the requirement of Regulation 522.6.204 option (i)

British Standard	Description of cable
BS 5467	Armoured cables having thermosetting insulation
BS 6724	Armoured cables having thermosetting insulation and low emission of smoke and corrosive gases when affected by fire
BS 7846	Armoured fire-resistant cables having thermosetting insulation and low emission of smoke and corrosive gases when affected by fire
BS EN 60702-1	Mineral insulated cables with a rated voltage not exceeding 750 V
BS 8436	300 / 500 V screened cables having low emission of smoke and corrosive gases when affected by fire, for use in thin partitions and building voids

Note: In order to be used to satisfy condition (i), the metallic covering of the cables listed has to be earthed.The metallic covering must also meet the requirements of Regulation 543.2.5 for a circuit protectiveconductor of the circuit concerned (that is, be suitable for that purpose and satisfy the requirements of indents (i) and (ii) of Regulation 543.2.4, relating to electrical continuity and cross-sectional area respectively).

Where, for example, a mineral insulated copper sheathed (MICS) cable or steel wire armoured (SWA) cable is used, the designer of the installation should confirm that the earthed metallic covering or armouring of the cable meets the requirements for a protective conductor for the circuit concerned. Regulation 543.2.5 permits the use of the metallic covering of a cable as a protective conductor provided it is suitable for the purpose and satisfies the requirements of items (i) and (ii) of Regulation 543.2.2, relating to electrical continuity and cross-sectional area, respectively.

Where screened cable to *BS 8436: 2011* is used, the requirements contained in Section 12 of that standard for the selection of protective devices and for the maximum current rating of cables must be met in addition to the requirements of *BS 7671*. These additional requirements are summarised below.

Only type B circuit-breakers to *BS EN 60898-1* or type B RCBOs to *BS EN 61009-1* may be used to provide overcurrent protection. Furthermore, the protective device must have the following characteristics given in Table 10.6.

Table 10.6 Requirements for circuit-breakers and RCBOs protecting *BS 8436* cables

Conductor size (mm^2)	Maximum let-through energy, I^2t (A^2t)	Maximum* rated current (I_n) of type B circuit-breaker or type B RCBO
1	42 000	16
1.5	42 000	20
2.5 (used as part of ring circuit wiring)	60 000	32
4	60 000	32

*A lower rating may be required to meet the requirements of *BS 7671* for electric shock or overcurrent protection.

For a protective device installed within a consumer unit complying with *BS EN 61439-3* its conditional rating can be considered to be 16 kA. Where a protective device is installed within an enclosure not complying with *BS EN 61439-3*, both its maximum breaking capacity and its energy class would need to be considered in more detail.

Regulation 522.6.204 Options (ii) & (iii) cables enclosed in earthed metallic conduit or trunking
In principle, an earthed steel conduit complying with *BS EN 61386-21* as described in condition (ii) or an earthed steel trunking or ducting complying with *BS EN 50085-2-1* as described in condition (iii) utilises the same method of protection as a cable with an earthed metallic covering and, by construction, provides a higher degree of protection against mechanical damage.

Regulation 522.6.204 Option (iv) the provision of mechanical protection
Option (iv) of Regulation 522.6.204 (shown in Table 10.4) relates to the provision of additional mechanical protection of a cable sufficient to prevent penetration of the cable by nails, screws and the like. It should be noted that this condition may prove hard to achieve in practice. Regulation 522.6.204 does not require such additional mechanical protection to be earthed.

A protective plate provided to prevent penetration by such fixings (see Fig 10.1) should be at least 3 mm thick. In situations where shot fired fixings are likely to be used, even a 3 mm plate will not prevent penetration and where shot fired fixings are being employed, the use of a metal plate would not be considered to afford compliance with Regulation 522.6.204 and would therefore be unacceptable. In such a case, cables should preferably be at least 50 mm from the surface of the joist or batten. If this is not practicable, then cables should either incorporate an earthed armour or metal sheath suitable for use as a protective conductor or be enclosed in earthed steel conduit, trunking or ducting.

Some proprietary metal plates intended for fixing to the surface of a joist or batten where a cable passes through are unsuitable for such use because they are too thin to prevent a nail, screw or the like penetrating the cable, or even to make the person driving in the fixing aware of the metal plate.

Regulation 522.6.204 Option (v) the use of SELV or PELV meeting the requirement of Regulation Group 414.4

Although *BS 7671* does not contain any regulations dealing specifically with the protection of extra-low voltage cables concealed in walls or flooring against impact, guidance is provided by Regulation 522.6.1. This regulation requires that all wiring systems shall be selected and erected so as to minimise damage from impact, abrasion, penetration, tension or compression during installation use or maintenance. Consequently, when using SELV or PELV to satisfy the requirements of Regulation 522.6.204 the wiring must be installed in a manner that minimises the risk of mechanical damage and satisfies the requirements of Regulation 522.6.1.

10.5.7 Protection of concealed cables in floors and ceilings

Regulation 522.6.201 requires that a cable installed under a floor or above a ceiling is run in such a position that it is not liable to be damaged by contact with the floor, the ceiling or any of their fixings. The regulation also requires that where a cable has to pass through a timber joist within a floor or ceiling construction, or through a ceiling batten, the cable must either:

1. be at least 50 mm measured vertically from the top, or bottom, as appropriate, of the joist or batten,

   or
2. comply with Regulation 522.6.203 (as described in Section 10.5.6.).

Fig 10.1 illustrates some of the methods of transversing joists permitted by Regulation 522.6.201.

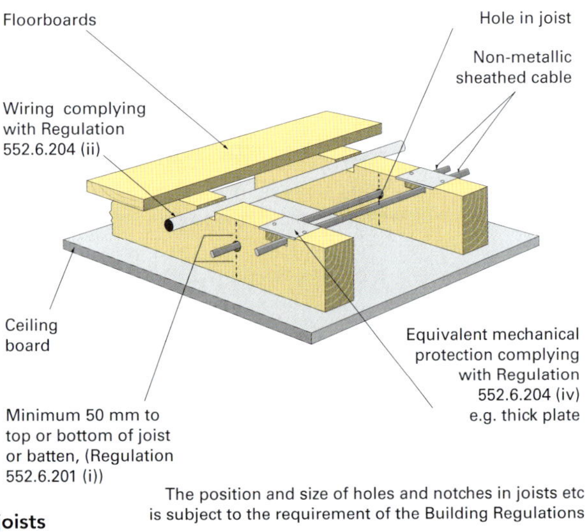

Fig 10.1 Cables in floor joists

Guidance on drilling holes and forming notches in ceiling and floor joists is included in Section 10.6.1 of this Guide

A cable meeting the 50 mm or greater requirement of Regulation 522.6.201 will, in normal circumstances, be at a sufficient distance from the top or bottom, as appropriate, of a joist or batten to prevent a nail, screw or the like penetrating the cable. However, if the installer knows or believes that the fixings being employed in a particular instance are likely to penetrate a cable placed at 50 mm from the top or bottom of the joist, then the cable must be placed at a depth within the surface sufficient to prevent the fixings from damaging the cable. If this is not practicable then at least one of requirements of 522.6.204 must be met (Regulation 522.6.201 refers).

10.5.8 Protection of concealed cables in walls

Cables embedded in a wall or partition can be vulnerable to penetration by nails, screws and the like, which can lead to the danger of electric shock or even fire. It is therefore important that cable runs are properly planned and that the cables are installed in such a manner as to afford compliance with the requirements of Regulation Group 522.6 dealing with protection against impact.

Where a wall or partition contains significant metal parts, such as would be found in walls and partitions constructed using a metal frame, there is a danger that as the result of damage to a cable all the metalwork within the wall or partition will become live. *BS 7671* treats walls or partition containing significant metal parts differently to walls and partition constructed using traditional methods such as blockwork or timber studwork which do not contain significant metal parts.

Walls with an internal construction NOT containing significant metallic parts

Regulations 522.6.202 deals with protection against impact where a cable is run at a depth of less than 50 mm from the surface of a wall or partition, where the internal construction does not contain significant metal parts. In this situation the cable must comply with the requirements of either of the following requirements:

(i) be installed in a prescribed zone (as shown in Fig 10.2) and be provided with additional protection by means of an RCD having the characteristics specified in Regulation 415.1.1, or

(ii) comply with Regulation 522.6.204 (as described in Section 10.5.6.).

Regulation 552.6.202 (i) describes the prescribed zones as:

a) within 150 mm from the top of the wall or partition, or
b) within 150 mm of an angle formed by two adjoining walls or partitions, or
c) either horizontally or vertically, to the point, accessory or switchgear on the surface of the wall or partition, to which the cable is connected.

Figure 10.2 provides a visual representation of these zones.

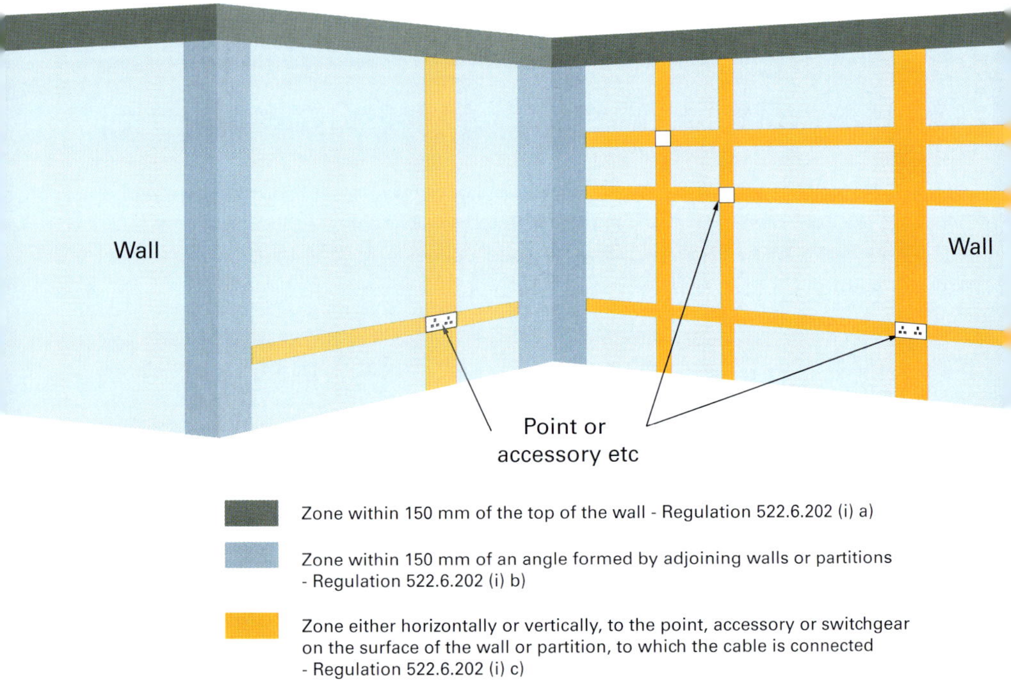

Fig 10.2 The zones detailed in Regulation 522.6.202 (i)

Generally, a zone created on one side of a wall or partition does not extend to the reverse side. However, where the location of the accessory, point or switchgear can be determined from the reverse side, a zone formed on one side of a wall or partition of 100 mm thickness or less extends to the reverse side. An example of this situation is shown in Fig 10.3.

The presence of a doorway in Fig 10.3 means that a person intending to drive a nail, screw or the like into the wall can reasonably determine the location of the accessory from the reverse side of the wall. A mirror image of the permitted cross-shaped zone created in the room with the accessory (room 1) is therefore considered to exist in the other room. Thus, the concealed wiring connected to the accessory is permitted to be within 50 mm of surface of the reverse side of the wall (just as it is permitted to be within 50 mm of the surface with the accessory), provided it is within the mirror image zone.

Apart from where the location of accessories, switchgear and the like can be determined from the reverse side, a cable that is concealed at a depth of less than 50 mm from either side of a wall or partition has to be in at least one of the permitted zones with respect to that side. The installation of a cable in such a zone, but passing through the box of an accessory rather than connecting to the accessory, is generally considered acceptable.

In order to comply with the requirements of Option (i) of Regulation 552.6.202 in addition to being run in an appropriate zone the wiring shall be provided with additional protection by means of an RCD having the characteristics specified in Regulation 415.1.1.

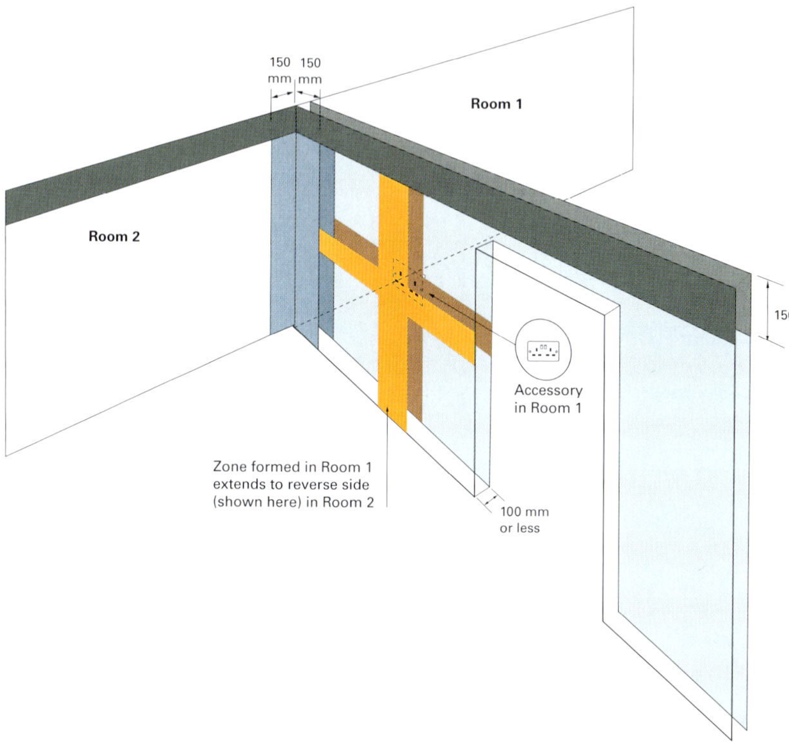

Fig 10.3 An example of where the location of an accessory can be determined from the reverse side of a wall

Walls with an internal construction containing significant metallic parts

Where a cable is concealed within a wall or partition where the internal construction contains significant metal parts, such as that shown in 10.4, Regulation 522.6.203 requires that irrespective of the depth of the cable from the surface of the wall or partition, the cable shall:

i) be provided with additional protection by means of an RCD having the characteristics specified in Regulation 415.1.1, or

ii) be mechanically protected sufficiently to avoid damage to the cable during construction of the wall or partition and during the installation of the cable, or

iii) comply with requirements of Regulation 522.6.204 (as described in Section 10.5.6.).

In addition, for a cable installed at a depth of 50 mm or less from the surface of a wall or partition containing significant metal parts the requirements of Regulation 522.6.202 previously described shall also apply.

Fig 10.4 The internal construction of a metal framed partition wall

10.5.9 Protection of cables embedded in plastered walls

Where cables are to be embedded in walls by 'wet' plaster, particular consideration also needs to be given to the requirements of Regulation 522.8.1, which requires a wiring system to be so selected and erected to minimise damage to the sheath and insulation during not only installation, but also during use and maintenance activities.

One obvious cause for concern when considering the installation of cables on walls that will subsequently be plastered is the possible damage to the cables that may be caused by the plasterer's trowel. A common method of protecting cables in such instances is to shield them with plastic or metallic capping, which also serves to secure the cables in place.

Whilst capping or indeed plastic conduit may be used to protect cables from damage during construction (see Fig 10.5), they do not provide adequate mechanical protection in service for cables located outside the zones described by Regulation 522.6.202.

Cables, concealed in plaster, may be installed outside these zones if they are protected by earthed metal conduit, as shown in Fig 10.6.

Fig 10.5 The use of cable capping

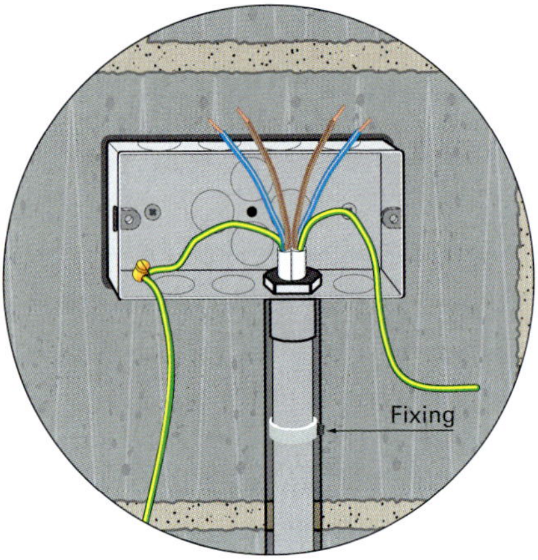

Fig 10.6 Earthed metal conduit

10.5.10 Other mechanical stresses

The following are classified in Appendix 5 of *BS 7671* as 'Other mechanical stresses' (external influence code 'AJ') and do not involve impact or vibration:

- abrasion (friction, chafing or scraping),
- compression (such as crushing),
- tension (pulling force), and
- torsion (twisting) stresses.

10.5.11 Damage caused during installation

Cables and other parts of a wiring system can easily be damaged during installation, and appropriate steps should be taken to prevent such damage. For example, the sheath of a cable may be torn or chafed where cables are being pulled through an opening in a joist or other part of a building structure, unless the opening is sufficiently large and smooth, and adequate care is exercised in the installation work. Similarly, cables already located in holes through joists may be damaged by the friction generated by other cables being pulled through the hole.

Care must be taken during the erection of conduit and trunking to ensure that there are no internal sharp edges. Internal burred metal must be removed at all joints, and bushes should be installed where cables pass through holes in enclosures.

To protect cables from damage any containment system employed should be adequately sized and there should be adequate means of access for drawing cables in and out.

Steps should also be taken to protect cables and other parts of a wiring system from damage due to the operations of other trades, such as plastering, the boarding of ceilings, floors and walls, or 'hot' work, such as welding.

In any event, wiring systems should be protected against impact, including that which could result from the operations of other trades.

10.5.12 Means of access to withdrawable wiring systems

For a wiring system designed to be withdrawable, such as cables in a conduit or ducting system, Regulation 522.8.6 requires adequate means of access to be provided for drawing cables in and out.

In addition, where a withdrawable wiring system is buried in the structure, Regulation 522.8.2 requires the wiring system for each circuit to be completely erected between access points **before** the cables are drawn in.

The above requirements are intended to avoid damage to cables when being drawn either in or out, due to inadequate means of access. For example, cable insulation could be damaged by abrasion, or due to excessive pulling force being applied to a cable.Excessive pulling force may also stretch the conductor of a cable thereby reducing its cross-sectional area and lead to 'hot spots', or cause one or more strands of the conductor to fracture, resulting in intermittent or permanent discontinuity.

10.5.13 Bending radii

Regulation 522.8.3 requires the radius of every bend in a wiring system to be such that the conductors or cables are not damaged.

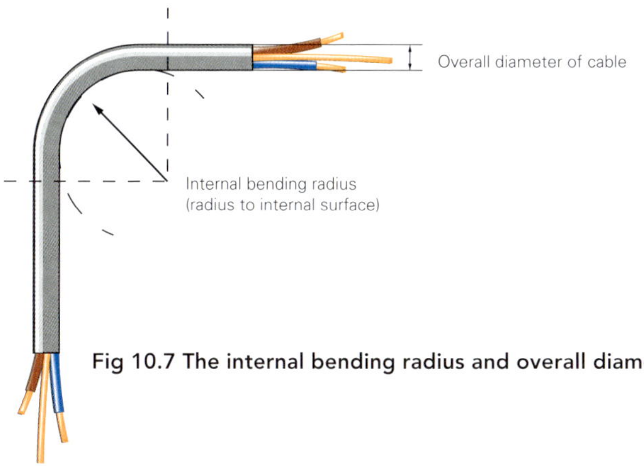

Fig 10.7 The internal bending radius and overall diameter of a cable

To comply with this requirement, a cable must not be bent to a radius smaller than a given value, based on its overall diameter (see Fig 10.7). To do so may cause damage to the conductors and insulation, or to the sheath, armouring or serving (if any) of the cable during erection of the installation or at some other time, such as under fault conditions.

Regulation 522.8.3 requires the radius of every bend in a wiring system to be such that the conductors or cables are not damaged. Regulation 522.8.3 also requires that no undue stress is placed upon terminations when cables or conductors are bent.

Table 10.7 Minimum internal radii for non-armoured cables

Cable type	Conductor material and construction	Cable overall diameter*	Factor to be applied to overall diameter of cable to determine minimum internal radius of bend	
		Non-armoured	Single strand conductor (solid)	Multi-strand conductor
Thermoplastic (PVC) or thermosetting (XLPE or rubber)	Copper and aluminium circular and circular stranded conductors	Not exceeding 10 mm	3	2
		Exceeding 10 mm	4	3
		Exceeding 25 mm	6	6
	Copper shaped conductors and solid aluminium	All	8	8

*For flat cables and diameter refers to the major axis.

10.5.14 Flexible wiring systems

Flexible wiring systems are sometimes used for certain applications in domestic electrical installations, for example, under-cupboard and wardrobe lighting, and final connections to items of equipment that are subject to vibration or which have to be moveable (such as for maintenance).

Regulation 521.9.2 requires that where equipment is intended to be moved during use, connection should be made using flexible cables.

Where an item of stationary equipment, such as a cooker, needs to be moved temporarily on occasion to permit connection, cleaning, maintenance or the like, the connection to the fixed wiring may be made by non-flexible cables (for example 'twin and earth' type cable). However, if the equipment is subject to vibration in use, flexible cable must be used for the connection (Regulation 521.9.3 refers).

10.5.15 Presence of fauna, flora and / or mould growth

Where exposed to conditions of fauna or flora and / or mould growth that may constitute a hazard, wiring systems and other equipment of an installation are required to be appropriately selected, or special measures adopted. (Regulation Groups 512.2, 522.9 and 522.10 of *BS 7671* refer).The term 'fauna' (animal life) includes insects, small animals and birds. Examples of flora (plant life) include plants, trees and other forms of vegetation; mould also has to be considered. Many electrical installations in external locations, as well as in some internal locations (such as greenhouses), may be exposed to fauna or flora and / or mould growth that could impair safety or functionality.

As with all external influences, it is best to avoid installing electrical equipment where it will be exposed to adverse effects. For example, distribution equipment (such as switchgear and consumer units) supplying outdoor installations should, where practicable, be installed in an indoor location.

If not suitably selected or adequately protected, equipment enclosures installed in contact with the ground (or buried in it) might become infested by insects.

Luminaires and other equipment designed to be installed outdoors may not necessarily prevent the ingress of insects and should be used only in accordance with the manufacturer's instructions. Cable joints for buried cables are usually compound filled and so ingress by fauna is not possible.

Equipment enclosures provided with small holes to allow collected or condensed water to drain away should not be installed where they are liable to become infested by insects.

Damage to cables from the gnawing of rats, mice and squirrels etc. is a relatively common occurrence. There is no evidence to show that these rodents are uniquely attracted to one type of cable. All types of cable, for example thermoplastic (PVC), rubber and lead sheathed and even steel wire armoured cables, seem susceptible to attack, resulting in exposure of the conductors.

Where conductors exposed by the gnawing action are surrounded by rodents' nesting materials, which are generally combustible, there is a risk of fire. This is in addition to the obvious danger of electric shock to persons from direct contact with an exposed live conductor.

Regulation 522.10.1 requires wiring systems to be suitably resistant to damage where the presence or actions of fauna constitutes a hazard and where not suitably resistant to such effects of itself, additional protective measures should be provided to counteract adverse effects. Such measures could include:

- choice of siting / location, or
- the provision of additional protection against mechanical damage either generally or locally, or
- a combination of the above.

Therefore, where cables are installed in areas that are likely to be, inhabited by rodents, such as might be found in the roof voids and cellars of certain older properties, the wiring system must be capable of resisting damage caused by gnawing.

It is advisable to employ cables or wiring systems such as the following for use in such areas:

- steel conduit, or trunking,
- mineral-insulated, metal-sheathed cables,
- steel wire armoured cables,
- thermoplastic (PVC) sheathed cables having galvanized steel braid covering,
- steel capping to protect short exposed runs of thermoplastic (PVC) insulated and sheathed cables.

Cables are less likely to be attacked by rodents where they are attached as far up a wall as possible, or, where installed in a roof void, fixed to the underside of the roof structure.

The enclosures of electrical switchgear, accessories and wiring systems (such as trunking) should be free of any openings through which rodents could gain entry to build nests and cause damage to cables or other equipment.

Cable entries to equipment, such as luminaires, must be effected through cable glands where necessary.

10.5.16 Solar and ultraviolet radiation

BS 7671 requires wiring systems that will be exposed to solar radiation (external influence code 'AN') to be either selected so that they are resistant to the damage from this external influence or protected from such damage (Regulations 522.2.1 and 522.11.1 refer).
Ultraviolet (UV) radiation can affect cable sheaths in two ways:

- degradation (splitting, etc), which may lead to danger, and
- colour fade can occur without the loss of mechanical properties.

Black cable sheaths made of thermoplastic (PVC) or rubber (but not natural rubber) material generally contain carbon black, an ingredient that gives good resistance to both degradation and colour fade. Grey cable sheaths of such materials contain less carbon black and are therefore likely to afford less resistance to degradation and colour fade than black sheaths.

Fig 10.8 Solar radiation

Carbon black is not present in thermoplastic or rubber sheaths of colours other than black or grey but an alternative UV stabiliser is sometimes incorporated.

Black or grey low-smoke zero-halogen sheaths used for armoured cables generally contain a UV stabiliser which resists degradation and colour fade.

Where cables intended mainly for use inside buildings incorporate a low-smoke zero-halogen sheath (such as for some types of cable commonly used for fire alarm systems), the sheath is generally not UV-stabilised as standard. However, a stabiliser may be incorporated to special order. A white sheath of such a cable will normally be materially stabilised but not colour stabilised.

It is important to appreciate that cable product specifications do not stipulate that cable sheaths should be resistant to UV radiation. Therefore, where cables are to be exposed to UV radiation, it is important that the electrical installation designer confirms that the sheaths are suitable for that purpose. As an alternative to using cables having a sheath that is resistant to UV radiation, cables may be protected by shielding them from direct sunlight.

10.5.17 Solar gain

Solar gain effectively increases the ambient temperature. As with any increase in temperature, the effective increase in temperature due to solar gain reduces the current-carrying capacity of cables. It may therefore be necessary to use cables having either an increased conductor size or a higher rated temperature in order to provide an adequate current-carrying capacity for the needs of the installation.

The ambient temperature rating factor (C_a) used when determining the current-carrying capacity of the cables should typically be based upon 10 °C above the maximum ambient air temperature in the UK unless the cables have been shielded from direct sunlight.

Installations in locations with glass roofs (such as greenhouses and conservatories) may also be subject to increased effective ambient temperature as a result of solar gain.

10.6 Impact upon the building structure

When chasing walls or drilling holes through the structural components of a dwelling electrical contractors need to exercise care to ensure that structural integrity of the dwelling is not compromised. Failure to take the necessary care can result in the electrical contractor being liable for expensive repairs.

10.6.1 Penetration of elements of building construction by wiring systems

A cable, trunking or cable tray sometimes has to pass through an element of building construction, such as a floor, wall or beam, necessitating the formation of a notch or opening in the element. To comply with Regulation 522.8.14, no wiring system should penetrate an element of construction that is intended to be load bearing unless the load bearing ability of the element remains satisfactory after such penetration.

For reasons of structural integrity, there may be restrictions on the dimensions or location of a notch or opening that may be formed in a load bearing element of construction. In some cases the formation of a notch or opening may be completely unacceptable. Where necessary, the advice of a suitably competent person should be obtained.

When working on domestic property electrical contractors are likely to encounter a variety of types of floor and ceiling joists, these may range from traditional solid timber joists to manufactured joists with timber or metal webs. Some pre-formed joists employ solid timber webs and when it is necessary to drill through this type of web it is most important to either follow the manufacturer's instructions or to take advice from a suitably qualified structural engineer.

Approved Document A no longer provides guidance on the drilling of solid timber joists, instead guidance is provided by *BS EN 1995-1: 2004* (as amended). The Timber Research and Development Association is a useful source of information regarding the drilling and notching of timber structural elements **www.trada.co.uk**

It is generally accepted, within the construction industry, that when dealing with simply supported[1], solid timber joists not greater than 250 mm in depth, adherence to the following guidance means that the effect of notches and holes need not be calculated (see Fig 10.10).

Notches should:

- be no deeper than 0.125 times the depth of a joist; and
- not be cut closer to the support than 0.07 of the span, nor further away than 0.25 times the span.

Holes should be:

- no greater diameter than 0.25 times the depth of the joist; should be drilled at the neutral axis[2]; and
- not less than 3 diameters (centre-to-centre) apart; and
- located between 0.25 and 0.4 times the span from the support.

It should be noted that notches or holes should not be cut in roof rafters, other than at supports where the rafter may be birdsmouthed[3] to a depth not exceeding 0.33 times the rafter depth.

Joists with solid timber web

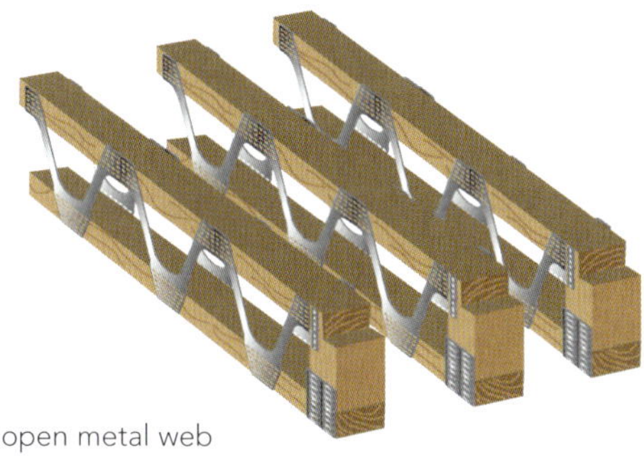

Joists with open metal web

Fig 10.9 Examples of prefabricated joists

[1] Simply supported means that each end of a floor or roof joist is supported.

[2] The neutral axis of a joist lies along is the centre line of a joist and is where the joist has neither a compressive or tensional stress.

[3] A woodworking joint generally used to connect a roof rafter to the top plate of a supporting wall.

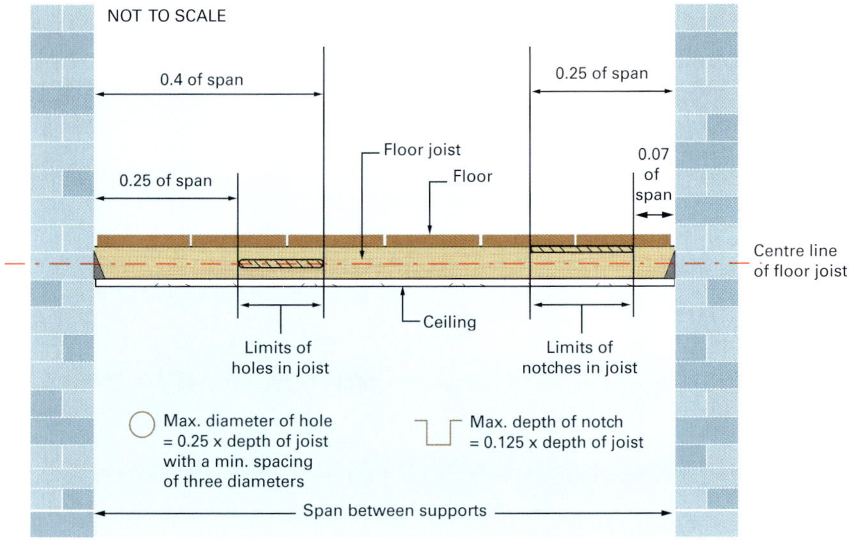

Fig 10.10 Acceptable positions for holes and notches in solid timber joists

10.6.2 Chases in walls

Typical building regulation guidance is as follows:

Vertical chases should not be deeper than ⅓ of the wall thickness or, in cavity walls, ⅓ of the thickness of the leaf.

Horizontal chases should not be deeper than ⅙ of the thickness of the wall or leaf.

Chases should not be so positioned as to impair the stability of the wall, particularly where hollow blocks are used.

10.7 Connections

10.7.1 General

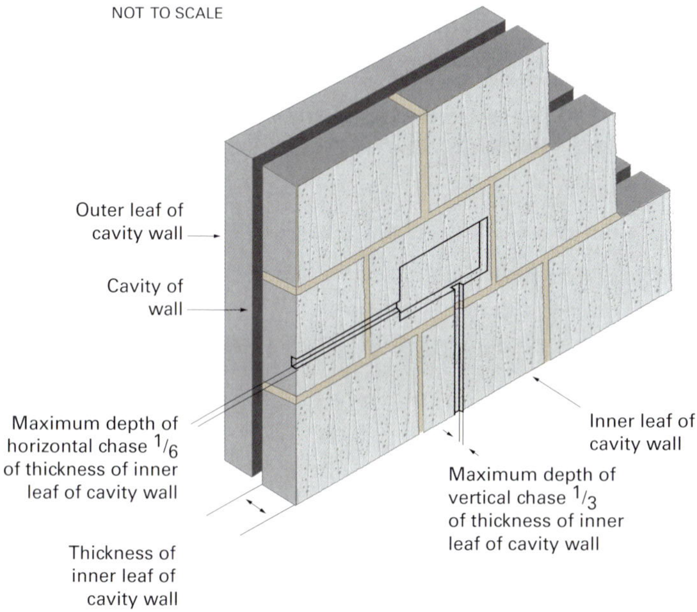

Fig 10.11 Examples of chases in the leaf of a cavity wall

The requirements relating to connections apply regardless of the voltage and therefore include extra-low voltage circuits. *BS 7671* contains a number of requirements concerning the construction of electrical connections, to protect against the following dangers which could arise from a poorly constructed connection.

- **Fire or other harmful thermal effects** (due, for example, to excessive temperature developed in a high resistance connection when it carries current, or to arcs or high temperature particles emitted from a loose connection)
- **Contact of a person or livestock with a live conductor** at a connection (direct contact). Basic protection is provided to prevent such a contact.
- **Fault conditions** (due, for example, to a high resistance or open circuit connection in a protective conductor causing the earth fault loop impedance of a circuit to exceed the maximum value required for automatic disconnection of the supply in the event of an earth fault).

10.7.2 Fundamental requirements

Every connection between conductors and between a conductor and equipment (such as switchgear, accessories or current-using equipment) is required to provide durable electrical continuity and adequate mechanical strength (Regulation 526.1 refers). These requirements are likely to be met by complying with the recommendations of this Guide.

Measures for protection against electric shock also have to be provided for connections in live conductors. Additionally, every connection in a live conductor or combined protective and neutral (PEN) conductor must be enclosed.

10.7.3 Selection of means of connection and termination

A means of connection between conductors, or between a conductor and an item of equipment, may take one of many forms, depending on the requirements of a particular installation. Examples of means of connection include a terminal of an accessory or item of switchgear, a terminal of a joint box, a connector block, a soldered or compression-type ferrule, and a soldered or compression-type lug bolted to a busbar.

Whatever means of connection is employed, Regulation 526.2 requires its selection to take account, as appropriate, of:

- (i) the material of the conductor and its insulation,
- (ii) the number and shape of the wires forming the conductor,
- (iii) the cross-sectional area of the conductor,
- (iv) the number of conductors to be connected together,
- (v) the temperature attained by the terminals in normal service such that the effectiveness of the insulation of the conductors connected to them is not impaired,
- (vi) where a soldered connection is used, the design must take account of creep, mechanical stress and temperature rise under fault current conditions (the term 'creep' refers to a gradual change of shape of material under stress.) and
- (vii) the provision of adequate locking arrangements in situations subject to vibration or thermal cycling.

In all cases, the limitations of equipment use specified in manufacturer's instructions must be observed.

(i) Material of the conductor and its insulation

The material of a conductor and its insulation can influence the selection of a means of connection in various ways, some examples of which are as follows.

Precautions must be taken to prevent galvanic (electrolytic) corrosion where certain dissimilar metals are placed in contact with each other. For example:

- an aluminium conductor must not be placed in contact with a terminal of brass or other metal having a high copper content, unless the terminal is suitably plated (with zinc or cadmium, for example) or other suitable precautions, are taken to prevent corrosion (such as fitting a suitable ferrule to the conductor)
- where an aluminium conductor and a copper conductor are joined together, precautions (such as plating) must be taken against corrosion.

Where an aluminium conductor is to be connected to a terminal, ferrule or lug, precautions must be taken to ensure that no undue mechanical pressure can be applied to the conductor, for example by the over-tightening of a clamping screw.

Crimped ferrules or lugs should always be correctly selected for the particular cable size and conductor shape and should be compressed only with the dies and nests recommended by the manufacturer.

Conductor insulation must not be subjected to temperatures exceeding its rated value in normal service. Particular attention is needed where conductors insulated with materials of different temperature ratings, such as thermoplastic (PVC) and thermosetting (XLPE or rubber), are to be connected. For example, the rated temperature of the insulation at the joint must be not less than the highest of those for the conductor insulation, or the conductors must be sized such that their normal operating temperatures will not exceed the lowest rated temperature of any of the insulating materials.

(ii) Number and shape of the wires forming the conductor

There are a wide range of conductor types which may need to be connected to each other or to equipment. These include (amongst others), conductors of cables (single-stranded, multi-stranded, flexible or non-flexible), braided conductors, circular rods (such as earth electrodes), and rectangular bars (such as busbars).

Whatever the number or shape of wires forming a conductor, the means of connection must be selected so as to:

- securely contain and anchor all the wires of the conductor, and
- make good electrical contact with the conductor.

The means of connection must not weaken or damage the conductors.

(iii) Cross-sectional area of the conductor

The means of connection must be suitable for the cross-sectional area of the conductor so that:

- adequate mechanical strength is provided by the means of connection (for example, to withstand the forces of thermal movement of the conductor),
- the conductor can be inserted into the terminal or other means of connection without using unreasonable force, and
- good electrical contact is provided between the conductor and the means of connection (for example, the terminal or other means of connection must be neither too large nor too small for the conductor, and must be correctly designed for the shape of the conductor).

Where it is necessary to reduce the cross-sectional area, or change the shape, of a conductor at the point of connection, this should be done by means of a recognised technique, such as the use of an appropriate pin connector.

(iv) The number of conductors to be connected together

BS 7671 does not place a general limit on the number of conductors that an individual terminal or other means of connection may be used to connect together. However, the number of conductors must be limited where necessary to facilitate the proper identification, inspection, testing and maintenance of the connections or conductors (Regulations 514.1.2 and 513.1 refer). Due attention should also be paid to appropriate product standards and manufacturers' instructions as required by Regulations 511.1 and 134.1.1.

(v) The temperature attained by the terminals in normal service

For most types of equipment, the temperature attained by terminals in normal service is unlikely to impair the effectiveness of the insulation of the conductors connected to them.

(vi) Soldered connections

Soldering of lugs and ferrules to conductors, although rarely carried out nowadays, was common practice until the use of compression-type sockets became established in the 1960s. In the absence of more precise information, soldered connections or joints are limited to a maximum temperature under fault conditions of 160 °C. They are therefore unsuitable for cable installations taking full advantage of the limiting final temperatures of above 160 °C offered by cables insulated with certain materials (such as thermosetting material).

Fitting and soldering of lugs and ferrules is a skilled practice and therefore relies on a jointer's skill to produce a solid ferrule or lug without leaving cavities or poorly tinned conductor strands. Such defects can result in a high resistance connection, leading to premature failure and the risk of fire.

(vii) Provision of adequate locking arrangements

Locknuts or other suitable locking arrangements must be provided to prevent terminal screws or other means of connection becoming loose in service where subject to vibration or thermal cycling.

External influences

Like any other parts of a wiring system, electrical connections must be appropriately selected and erected to be suitable for the external influences they are likely to encounter, such as the following, which are referred to in Section 522 of *BS 7671*.

- Ambient temperature (taking account of the highest and lowest local ambient temperatures).
- External heat sources (such as solar gain).
- Presence of water, high humidity or solid foreign bodies (prevention of ingress and / or accumulation, as applicable).
- Presence of corrosive or polluting substances.
- Impact.
- Other mechanical stresses (for example, where necessary to prevent damage due to snagging, connection boxes should be securely fixed, and means such as a cable clamp should be provided to prevent strain on terminations).
- Presence of flora and / or mould growth or fauna.
- Radiation (solar or ultraviolet).
- Building design (for example structural movement).

See also Section 10.5 of this Guide.

10.7.4 Enclosure of connections

Irrespective of nominal voltage, every termination and joint in a live conductor or PEN conductor is required to be contained within an enclosure selected in accordance with Regulation 526.5. This includes connections in extra-low voltage circuits such as those supplying extra-low voltage luminaires.

The purpose of such an enclosure is to protect persons, and property against harmful thermal effects, such as arcs or high temperature particles which may be emitted from a loose connection (Regulation 421.1.3 refers). The enclosure may also provide basic protection against electric shock and protection against external influences.

10.7.5 Protection against electric shock

As for any other part of an electrical installation, a connection in a live (line or neutral) conductor must be provided with a measure of basic protection and a measure of fault protection.

Basic protection

The enclosure may often be used for basic protection. This is provided that the design of the enclosure meets the requirements of *BS 7671* for such use, such as affording an appropriate degree of protection (IP Code), and having suitable facilities (such as a screw-fixed cover) where opening of the enclosure is necessary.

Where the enclosure is not relied on for basic protection, the live parts at a connection must normally be insulated or placed behind a barrier, in accordance with the applicable requirements. It should be noted that the use of insulating tape will **not** meet the requirements of *BS 7671* for basic protection.

Fault protection

At least one of the protective measures referred to in Regulation 410.3.3 must be provided for an electrical connection (such as Automatic Disconnection of Supply, which includes the earthing of exposed-conductive-parts of metal enclosures).

In addition, the electrical continuity of connections in protective conductors is an essential part of ensuring that the earth fault loop impedance of a circuit does not exceed the maximum value required for automatic disconnection of the supply in the event of an earth fault, and is maintained at a suitably low value.

10.7.6 Accessibility of connections

Regulation 526.3 requires every connection (with certain exceptions, listed in the regulation) to be accessible for inspection, testing and maintenance.

Except where a concealed conduit, trunking, metal armouring or metal cable sheath is used as a circuit protective conductor (cpc), the requirements of Regulation 526.3 for accessibility apply no less to a connection in a protective conductor than to a connection in any other conductor.

10.7.7 Protective conductor connections involving the metal sheath or armouring or metal enclosure

Subject to all the relevant requirements being met, a protective conductor may consist of a metal part of a wiring system, such as a metal sheath, armouring or screen of a cable, or a metal conduit or other enclosure or electrically continuous support system for conductors (see Regulation 543.2.2).

The electrical continuity of connections involving such parts of a wiring system is required (no less than that of the parts themselves) to be assured in such a way as to be protected against mechanical damage, chemical and electrochemical deterioration.

10.8 Cable supports

10.8.1 General

The basic requirements for the support of cables are given in Regulations 522.8.4 and 522.8.5 of *BS 7671*. In particular, cables should be supported in such a way that they will not be exposed to undue mechanical strain, including any strain caused by their own weight, and that there is no appreciable mechanical strain on the terminations of the conductors. Where a cable is supported at intervals, rather than continuously, the intervals should be appropriate to meet the above requirements for avoidance of strain.

BS 7671 does not stipulate the particular means by which cables should be supported, but leaves this to be dictated by the needs of the installation concerned. However, a number of essential requirements need to be met, some examples of which are expanded upon below.

10.8.2 Surfaces in contact with a cable

All surfaces in contact with a cable must be reasonably smooth and free from sharp corners or edges likely to damage the cable (Regulation 522.8.11 refers). This applies to cable clips, parts of a wiring system (such as conduit and cable tray) and parts of a building (such as walls, ceilings and floors).

10.8.3 Suitability for external influences

Cable supports and their fixings, like any other parts of a wiring system, must be selected so as to be suitable for the external influences they are likely to be exposed to at their location (Section 522). For example, where vibration is likely to occur, locking arrangements such as shakeproof washers may be required.

10.8.4 Avoidance of electrolytic corrosion

Metallic cable clips, support systems (such as cable tray), cable sheaths and armour etc. of dissimilar metals between which electrolytic corrosion is liable to occur, should not be placed in contact with each other (Regulation 522.5.2 refers). Such corrosion is possible between, for example, zinc (as used in galvanizing) and copper, especially in the presence of moisture.

10.8.5 Enclosure of non-sheathed cables

Non-sheathed cables for fixed wiring are required by Regulation 521.10.1 to be enclosed in conduit, ducting or trunking. This requirement does not apply to a protective conductor complying with Section 543 of *BS 7671*. However, where the protective conductor belongs to an AC circuit installed in a ferrous enclosure (such as steel conduit, trunking or ducting) the protective conductor must be installed in the same enclosure in order to meet the requirements of Regulation 521.5.1 relating to electromagnetic effects.

Vertical runs of trunking or ducting exceeding 5 m in length

For cables of any type in a vertical run of trunking or ducting exceeding 5 m in length, intermediate cable supports should be provided.

Vertical runs of conduit exceeding 5 m in length

For non-sheathed cables in a vertical run of conduit exceeding 5 m in length, precautions (such as additional cable supports) should be taken to protect the cable insulation on the bend at the top of the run against undue compression.

Vertical runs of cable supported from the top

Where a vertical run of cable is supported from the top, such support should comprise a clip and a rounded support of radius not less than that recommended in Table 10.8. In addition, for reasons of protection against mechanical damage (Regulation Groups 522.6 and 522.8), the cable should be inaccessible and unlikely to be disturbed.

BS 5839 for fire detection and alarm systems, and *BS 5266* for emergency lighting systems may contain additional recommendations.

10.8.6 Cables resting without fixings

Cables resting without fixing on part of a building (such as on the top surface of a plasterboard ceiling beneath floorboards) should be located where they are inaccessible and unlikely to be disturbed.

10.8.7 Spacing between supports

BS 7671 requires the intervals between cable supports (as well as the means of support) to be such that cables do not suffer mechanical damage or strain. Table 10.8 gives suggested maximum spacings between supports for both horizontal and vertical runs of cable.

As *BS 7671* is principally concerned with the safety aspects of support systems, supports additional to those indicated in Table 10.8 are often needed to make the finished installation visually acceptable. Additional supports may also be needed in an installation subjected to vibration of medium or high severity, or where there is an increased risk of mechanical damage, such as by accidental snagging of the cable.

Special considerations sometimes apply to the spacing and arrangement of supports for distribution cables installed in free air, due to thermal expansion, and for the maximum lengths of span and minimum heights for overhead wiring linking buildings etc.

10.8.8 Wiring system supports

Table 10.8 Maximum spacing of supports for cables in accessible positions, the entire support being derived from the clips

Overall diameter of cable, Φ [3] (mm)	Non-armoured thermosetting cable [1]	
	Horizontal [2][4]	Vertical [2]
$0 < \Phi \leq 9$	250	400
$9 < \Phi \leq 15$	300	400
$15 < \Phi \leq 20$	350	450
$20 < \Phi \leq 40$	400	550

Notes:
1. The spacings given in this table are maximum values, and for good workmanship in certain circumstances the spacings may need to be reduced.
2. For the spacing of supports of cables having an overall diameter not exceeding 40 mm.
3. For flat cables, taken as the dimension of the major axis.
4. The spacings stated for horizontal runs may be applied also to runs at an angle of more than 30° from the vertical. For runs at an angle of 30° or less from the vertical, the vertical spacings are applicable.

In the event of a fire, the heat generated is likely to cause plastic cable ties and non-metallic conduit or trunking to fail, and this would result in surface run wiring systems in particular, hanging at low level. This has the potential to impede firefighters in the execution of their duties, and possibly hinder evacuation from the premises. Regulation 521.10.202 requires that all wiring systems shall be supported in such a way that, in the event of a fire, they will not be liable to premature collapse.

Premature collapse of wiring systems can be avoided in a number of ways including:

(a) metallic clips inside plastic trunking systems

(b) metallic clips holding the plastic trunking lid on

(c) metallic clips on wall-mounted cables

An example of typical cable supports are shown in Fig 10.12

Cables installed in floors or above ceilings or behind plasterboard are not considered to be an entanglement risk and do not require additional support.

Cables installed in steel conduit, steel trunking and on cable tray, ladder rack and the like, are also not considered to be an entanglement risk.

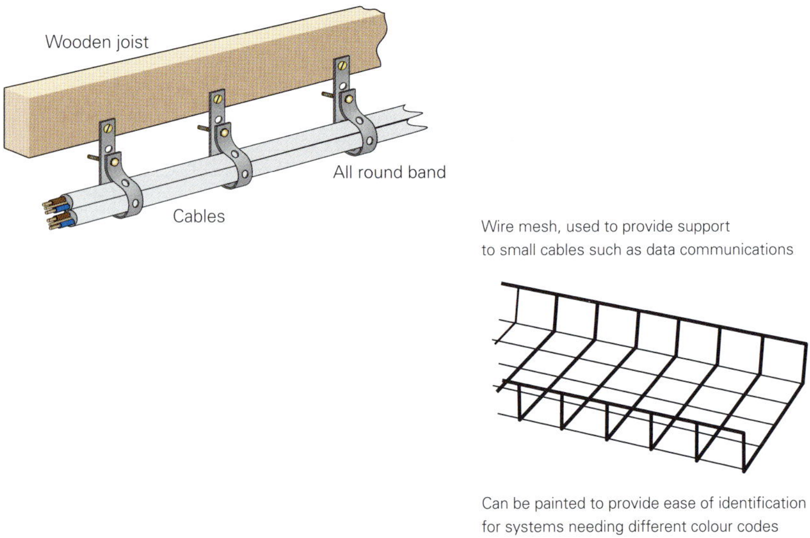

Fig 10.12 Typical examples of cable supports that would meet the requirements of Regulation 521.10.202

CHAPTER 11
SPECIAL INSTALLATIONS AND LOCATIONS

11.1 Introduction and scope

This Chapter looks at the requirements for the following special installations or locations falling within the scope of Part 7 of *BS 7671* that may be found in domestic premises.

- Locations containing a bath or shower (Section 701).
- Swimming pools (Section 702).
- Saunas (Section 703).
- Electric vehicle charging installations (Section 722).

Section 700 of *BS 7671* explains that the requirements within Part 7 supplement or modify the general requirements found elsewhere in that standard. In the absence of any additional or modified requirements in a Part 7, the general requirements apply unchanged.

A zonal concept is applied for some of the special locations covered in this chapter, whereby the requirements for safety are based on the perceived degree of risk of electric shock at a particular point within the location.

In this Chapter the term '30 mA RCD' is often used. This term means a Residual Current Device (RCD) with a rated residual operating current ($I_{\Delta n}$) not exceeding 30 mA. Such a device is recognised in AC systems as providing additional protection in the event of failure of the provision for basic protection and / or the provision for fault protection or carelessness by users (Regulation 415.1.1 refers).

The use of such devices is not recognised as a sole means of protection and does not obviate the need to apply one of the protective measures specified in Sections 411 to 414 of *BS 7671* (Regulation 415.1.2 refers).

11.2 Locations containing a bath or shower

Locations containing a bath, shower or cabinet containing a shower and / or bath and the surrounding areas are considered to be locations where there is an increased risk of electric shock due to:

- a reduction in body resistance caused either by bodily immersion or by wet skin, and
- likely contact of substantial areas of the body with earth potential.

Consequently, a number of specific requirements are contained within Section 701 (Locations containing a bath or shower) of *BS 7671*.

In addition to the requirements given in Section 701, all other relevant sections of *BS 7671* apply to such locations, albeit supplemented and modified by Section 701. It should be noted that the requirements of Section 701 must be applied where a bath or shower is located in a room having a different primary purpose, such as a bedroom.

11.2.1 Zones

Zone 0 as described in Regulation 701.32.2, is the interior of the bath tub or shower basin. For showers without a basin, zone 0 extends to 0.10 m vertically and to the same extent horizontally as zone 1 as described below.

Zone 1 is the three dimensional space limited by:

- The finished floor level and whichever is the higher, the horizontal plane aligning with the highest fixed shower head or water outlet or the horizontal plane lying 2.25 m above the finished floor level.
- The vertical surface:
  a) forming the edge of the bath tub or shower basin, or
  b) for showers without a basin, at a distance of 1.20 m from the centre point of the fixed water outlet on the wall or ceiling.

Zone 1 does not include zone 0.

The space under the bath tub or shower basin is considered to be zone 1 unless such space is accessible only with a tool, in which case it is considered to be outside the zones (Regulation 701.32.3 refers).

Zone 2 is the three dimensional space limited by:

- The finished floor level and whichever is the higher, the horizontal plane aligning with the highest fixed shower head or water outlet or the horizontal plane lying 2.25 m above the finished floor level.
- The vertical surface between the boundary of zone 1 and the parallel vertical surface at a distance of 0.60 m from the zone 1 border.

For showers without a basin, there is no zone 2 but zone 1 is increased, extending 1.20 m horizontally from the fixed water outlet.

The zones are limited by walls, doors, fixed partitions, ceilings and floors where these effectively limit the extent of a zone.

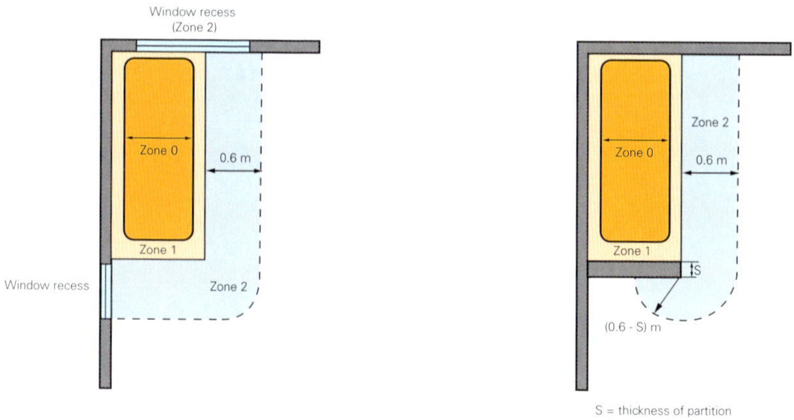

Fig 11.1 Bath tub; plan views (with and without a permanently fixed partition) to illustrate zone boundaries

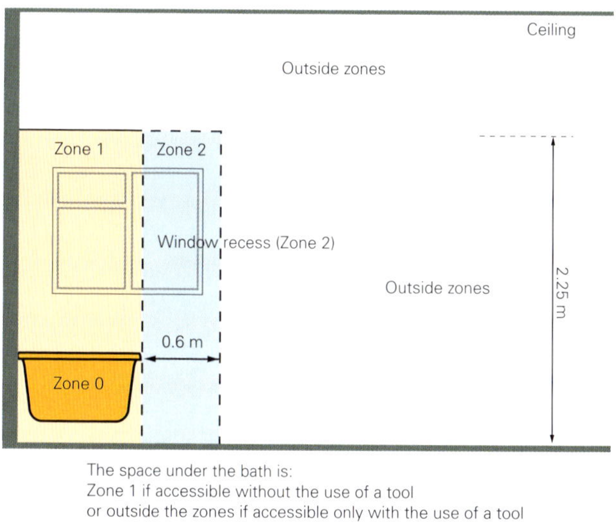

Fig 11.2 Bath tub; elevation

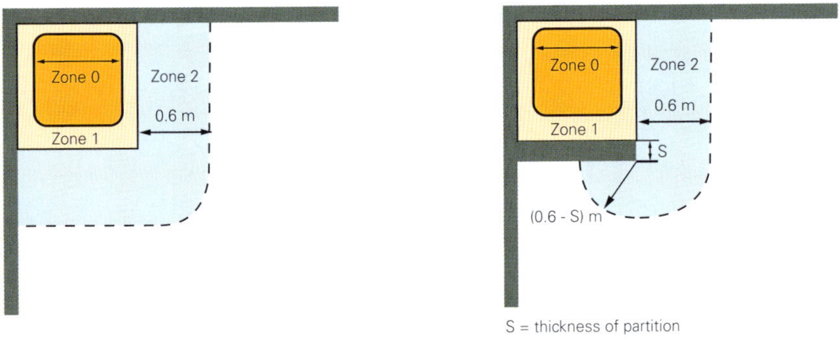

Fig 11.3 Shower basin; plan views (with and without a permanently fixed partition) to illustrate zone boundaries

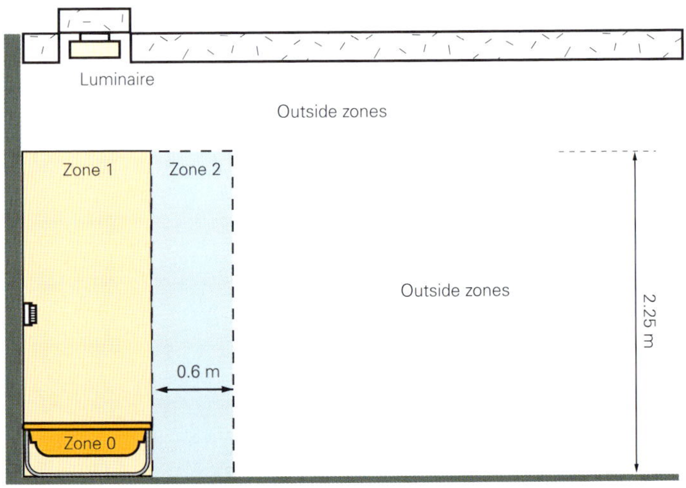

The space under the basin is:
Zone 1 if accessible without the use of a tool
or outside the zones if accessible only with the use of a tool

Fig 11.4 Shower basin; elevation

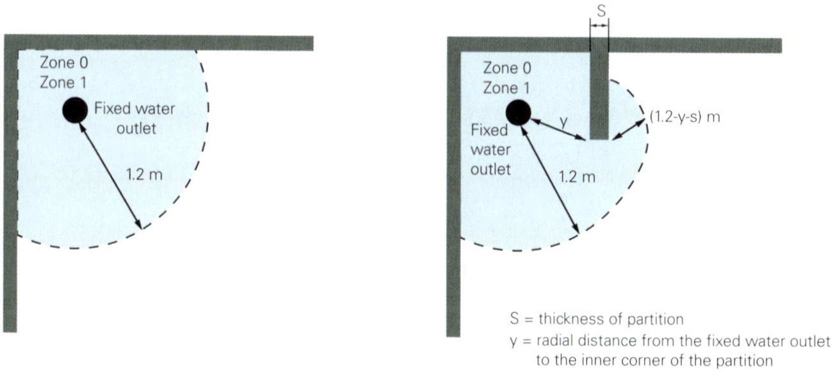

Fig 11.5 Shower without a basin; plan views (with and without a permanently fixed partition) having a fixed water outlet

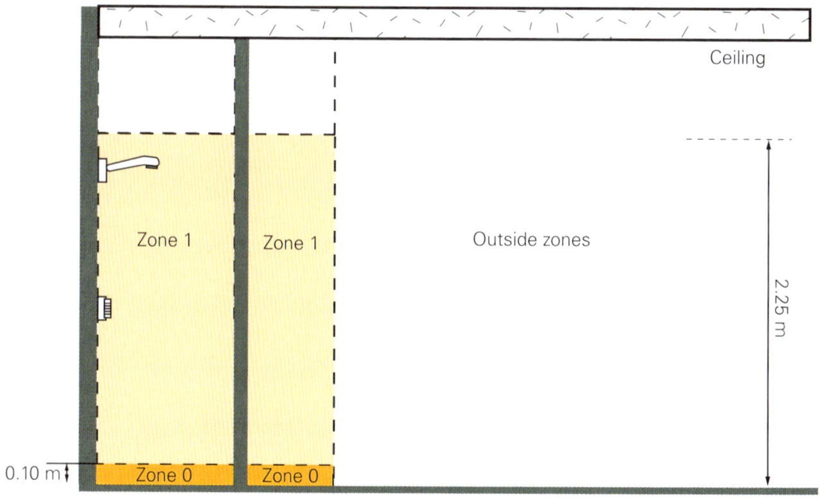

Fig 11.6 Shower without a basin; elevation with a permanently fixed partition

11.2.2 Protection against electric shock

BS 7671 does not allow basic protection to be provided by obstacles or protection by placing out of reach anywhere within in a location containing a bath or shower (Regulation 701.410.3.5 refers).

Further, fault protection by non-conducting location or by earth-free local equipotential bonding is similarly prohibited (Regulation 701.410.3.6 refers).

11.2.3 Additional protection

Regulation 701.411.3.3 requires that all low voltage circuits (typically 230 V) serving or passing through zones 1 or 2, but not serving the location containing a bath or shower, are required to have additional protection by one or more RCDs not exceeding 30 mA. This includes any circuits provided for lighting.

It is the responsibility of the installation designer to configure the consumer unit and the final circuits so that all circuits extending into the location have the necessary RCD protection. The possibility of unwanted tripping must be considered and, if necessary, suitable arrangements made (Regulation Group 314 and Regulation 531.3.2 refer). See also Section 3.5 of this Guide.

Where extra-low voltage by means of SELV or PELV is used, whatever the nominal voltage, basic protection for equipment in zones 0, 1 and 2 must be provided by basic insulation or barriers or enclosures affording a degree of protection of at least IPXXB or IP2X (Regulation 701.414.4.5 refers).

11.2.4 Supplementary equipotential bonding

The purpose of supplementary equipotential bonding is that under earth fault conditions, all exposed-conductive-parts and extraneous-conductive-parts should rise to the same potential. Where there is any doubt about the effectiveness of the supplementary bonding then the condition stated in Regulation 415.2.2 applies.

Regulation 701.415.2 states that where the location containing a bath or shower is in a building with a main protective equipotential bonding system (as should generally be the case), supplementary protective equipotential bonding may be omitted where all of the following conditions are met:

i) all final circuits of the location comply with the requirements for automatic disconnection according to Regulation 411.3.2, and

ii) all final circuits of the location have additional protection by means of a 30 mA RCD in accordance with Regulation 701.411.3.3, and

iii) all extraneous-conductive-parts of the location are effectively connected to the protective equipotential main bonding according to Regulation 411.3.1.2.

Note: The effectiveness of the connection of extraneous-conductive-parts in the location to the main earthing terminal may be assessed, where necessary, by the application of Regulation 415.2.2 regarding the value of resistance (Regulation 701.415.2 refers).

Where fitted, supplementary bonding connections, as with all electrical connections or joints, should be:

i) accessible for inspection, testing and maintenance (Regulation 526.3 refers) apart from connections or joints made in one of the ways exempted by that Regulation, and

ii) identified by a permanent label at the point of connection of every bonding conductor to an extraneous-conductive-part (Regulation 514.13.1 refers).

The supplementary bonding connection to a fixed electrical appliance, such as an electric towel rail, may be provided by the circuit protective conductor (cpc) within a short length of flexible cable connecting the appliance to an accessory having a flex outlet (Regulation 544.2.5 refers).

External influences

Each item of electrical equipment in a location containing a bath, shower, or cabinet containing a shower and / or bath, no less than in any other area, must be selected and erected so as to be suitable for the external influences likely to occur at the particular point of installation. Such influences are likely to include steam (leading to condensation), falling drops of water and sprays and / or jets from shower nozzles.

Shower units (incorporating an electrical instantaneous water heater), as with all equipment, must be suitable both for the zone and the conditions likely to occur.

Regulation 701.512.2 of Section 701 contains a number of requirements regarding the degree of protection against ingress of water into electrical equipment to be applied in addition to general requirements for external influences given in Regulation Group 512.2 and Section 522.

The minimum permitted degree of protection related to the zones is shown in Table 11.1.

Table 11.1 Minimum required degrees of ingress protection

Zone(s)	Minimum degree of protection(2) IP code
Zone 0	IPX7
Zones 1 and 2(1)	IPX4
Zones 1 and 2 where water jets are likely to be used for cleaning purposes in communal baths or communal showers	IPX5
Outside zones 0, 1 and 2	General rules (see note 2)

Note 1: Shaver supply units to *BS EN 61558-2-5* may be installed in zone 2 if located where direct spray from a shower is unlikely (Regulation 701.512.2 refers).
Note 2: Equipment has to be suitable for the conditions likely to occur at the particular point of installation. This may mean a higher IP rating than the minimum value required by the Table.

Switchgear, controlgear and accessories

Table 11.2 identifies the switchgear, controlgear and accessories that may be installed within a particular zone in the special location (Regulation 701.512.3 refers).

Table 11.2 Permitted equipment in the zones

Zone	Permitted equipment
Zone 0	Switches and controls incorporated in fixed current-using equipment suitable for use in the zone. Insulating pull cords of cord operated switches.
Zone 1	Switches and controls incorporated in fixed current-using equipment suitable for use in the zone. Switches of SELV circuits supplied at a nominal voltage not exceeding 12 V AC rms or 30 V ripple-free DC, the safety source being installed outside of zones 0, 1 and 2. Insulating pull cords of cord operated switches.
Zone 2	Switches and controls incorporated in fixed current-using equipment suitable for use in the zone. Switches or socket-outlets of SELV circuits, the safety source being installed outside of zones 0, 1 and 2. Shaver supply units complying with *BS EN 61558-2-5* (if located where direct spray from a shower is unlikely). Insulating pull cords of cord operated switches.
Outside zones 0, 1 and 2	The general requirements of *BS 7671* apply.

Note: Except for SELV socket-outlets complying with Section 414 and shaver supply units complying with *BS EN 61558-2-5*, socket-outlets are prohibited within a distance of 3 m horizontally from the boundary of zone 1.

A question sometimes arises of whether wall-mounted lighting switches and other such accessories are permitted in a location containing a bath or shower. Regulation 701.512.3 does not preclude the installation of such equipment outside of the zones.

However, the general requirements of *BS 7671* are applicable to locations containing a bath or shower and Regulation 512.2.1 calls for every item of equipment to be of a design appropriate to the situation in which it is to be used, or its mode of installation shall take account of the conditions likely to be encountered.

The requirements of *BS 7671* may be met by:

- installing a wall-mounted light switch with an IP rating suitable for its position within the location
- using a pull-cord switch complying with *BS EN 60669-1: 1999 (2008)* with an IP rating suitable for its position. The body of the switch must be installed outside of zones 0, 1 and 2 but the pull cord itself is permitted to enter zones 1 or 2.

Fixed current-using equipment

Table 11.3 identifies, with respect to the zone, the fixed current-using equipment that may be installed in a location containing a bath or shower (Regulation 701.55 refers).

Table 11.3 Fixed current-using equipment that may be installed in a location containing a bath or shower

Zone	Permitted equipment
Zone 0	Current-using equipment that: • complies with the relevant standard and • is suitable for the conditions of the zone according to the manufacturer's instructions for use and mounting, and • is fixed and permanently connected and • is protected by SELV at a nominal voltage not exceeding 12 V AC rms or 30 V ripple-free DC and • the source is installed outside of zones 0, 1 and 2.
Zone 1	The following fixed, permanently connected current-using equipment provided it is suitable for installation in zone 1 according to the manufacturer's instructions: • whirlpool units • electric showers • shower pumps • equipment protected by SELV or PELV at a nominal voltage not exceeding 25 V AC rms or 60 V ripple free DC the safety source being installed outside zones 0,1 and 2 • ventilation equipment • towel rails • water heating appliances • luminaires
Zone 2 and outside zones 0 and 1	The general requirements of *BS 7671* apply.

Low voltage socket-outlets

Low voltage socket-outlets, such as a 13 A, 230 V socket-outlet, are not permitted within a distance of 3 m horizontally from the boundary of zone 1 in a location containing a bath or shower (see Regulation 701.512.3 refers). The circuit supplying the socket-outlet, in common with all the low voltage circuits of the location, must have additional protection provided by an RCD not exceeding 30 mA (Regulation 701.411.3.3 refers).

Shaver supply units and SELV socket-outlets

Shaver supply units complying with *BS EN 61558-2-5* may be installed in zone 2 and beyond as can SELV socket-outlets complying with Section 414. Such shaver supply units incorporate a safety isolating transformer to afford protection by electrical separation.

Shaver supply units are not normally splash-proof and whilst Regulation 701.512.3 permits their installation outside zones 0 and 1 they should be positioned well away from baths or showers.

Normal lighting switch height (that is, between 450 mm and 1200 mm above finished floor level) will usually safeguard against splashing from a wash basin.

Stationary equipment

Washing machines, tumble dryers and the like may be installed beyond zone 2 in a location containing a bath or shower supplied by a socket-outlet or fused connection unit.

However, in extreme cases, a manufacturer's agent may not be prepared to carry out servicing or maintenance on an appliance in a location containing a bath or shower until it is moved to what they consider to be a safer location, even where all the relevant electrical installation requirements of *BS 7671* have been met.

Electric floor heating systems

Regulation 701.753 states that only heating cables according to relevant product standards or thin sheet flexible heating elements according to the relevant equipment standard should be employed for electric floor heating systems in a location containing a bath or shower.

Any such heating cable or element should either have either a metallic sheath, or a metallic enclosure or a fine mesh metallic grid which should be connected to the protective conductor of the supply circuit, except where the protective measure of SELV is provided for the floor heating system.

This Regulation further states that the protective measure of protection by electrical separation is not permitted for floor heating systems within the location.

Table 11.4 provides a summary of the requirements of *BS 7671* for locations containing a bath or shower.

Table 11.4 Summary of the requirements for a location containing a bath or shower

Zone	Outside Note 5	2 Note 6 IPX4 see (Note 9)	1 Note 6 IPX4 see (Note 9)	0 Note 6 IPX4 see (Note 9)
230 V wall mounted plate switch	✓	✓	✓	✓
Luminaire	✓	✓	✓ note 3	✗
SELV or PELV safety source	✓	✗	✗	✗
Shaver supply unit	✓	✓ note 2	✗	✗
Ventilation equipment	✓	✓	✓ note 3	✗
Pull cord switch mechanism	✓	✗	✗	✗
Insulated pull cord	✓	✓	✓	✓
SELV switches / socket-outlets	✓	✓ note 8	✓ note 7	✗
Switches or controls in fixed current-using equipment suitable for use in the zone	✓	✓	✓	✓
Whirlpool unit, electric shower, shower pump, ventilation equipment, towel rail, water heating appliance	✓	✓	✓ note 3	✗
Equipment, such as a fan or light, protected by SELV or PELV at a nominal voltage not exceeding 25 V AC rms or 60 V ripple-free DC	✓	✓	✓ note 3	✗
SELV fixed light or fan. Equipment protected by SELV at a nominal voltage not exceeding 12 V AC rms or 30 V ripple-free DC	✓	✓	✓	✓ note 4
230 V socket-outlet	✓ note 1	✗	✗	✗

Note 1 Prohibited within a distance of 3 m horizontally from the boundary of zone 1.

Note 2 A shaver supply unit must be to *BS EN 61558-2-5*. The requirement for a degree of protection of a minimum of IPX4 in zone 2 does not apply to shaver units situated where direct spray from showers is unlikely.

Note 3 Equipment must be fixed, permanently connected, suitable for zone 1 according to the manufacturer's instructions.

Note 4 Equipment must be fixed, permanently connected, and suitable for zone 0 according to the manufacturer's instructions.

Note 5 The general requirements in Parts 1 to 6 of the Regulations are applicable outside of the zones including Regulation 512.2.1 which requires equipment to be of a design appropriate to the situation in which it is to be used or its mode of installation must take account of the conditions likely to be encountered.

Note 6 The general requirements in Parts 1 to 6 of the Regulations are applicable within the zones of a location containing a bath or shower but are supplemented by the additional requirements in Part 701 of *BS 7671*.

Note 7 SELV switches must be supplied at a nominal voltage not exceeding 12 V AC rms or 30 V ripple-free DC and the safety source must be outside zones 0, 1 and 2.

Note 8 SELV switches and socket-outlets permitted but the safety source must be outside zones 0, 1 and 2.

Note 9 Equipment must have a minimum protection of IPX4 (IPX5 if water jets are likely).

Note 10 Equipment must have a minimum protection of IPX7.

It must be remembered that all circuits of the location must have additional protection provided by one or more RCDs not exceeding 30 mA (Regulation 701.411.3.3 refers).

Furthermore, providing all final circuits of the location have additional protection, meet the requirements for automatic disconnection and the installation is fitted with protective main equipotential bonding, then supplementary equipotential bonding may be omitted (Regulation 701.415.2 refers).

11.3 Swimming pools

Swimming pools, paddling pools and their surroundings are considered to be locations where there is an increased risk of electric shock due to:

- a reduction in body resistance caused either by bodily immersion or by wet skin, and
- likely contact of substantial areas of the body with earth potential.

Consequently, a number of specific requirements are contained within Section 702 (Swimming pools and other basins) of *BS 7671*. These should also be applied to the basins of any fountains if is intended or likely that they are liable be occupied by people.

The requirements for fountains not expected to be occupied by persons are not discussed in this Guide.

In addition to the requirements given in Section 702, all other relevant sections of *BS 7671* apply to such locations, albeit supplemented and modified by Section 702.

Early design collaboration and installation co-ordination are needed if significant problems are to be avoided during construction and verification of the electrical installation. Without such collaboration and co-ordination, problems may occur, for example, in relation to the installation of metallic grids in floors where necessary, the installation of supplementary bonding conductors, the positioning of electrical equipment relative to the zones referred to in this Guide, and the protection of electrical equipment against external influences.

Special requirements beyond those of *BS 7671* may apply to swimming pools for medical use, although this falls outside of the scope of this Guide.

11.3.1 Zones

Regulation 702.32 divides the basins of pools (and fountains intended for occupation by persons) and their surroundings into zones 0, 1 and 2 for the purposes of selecting suitable equipment and shock protection measures. The zones are explained in Table 11.5 and illustrated in Figures 11.7, and 11.8.

Table 11.5 Zone definitions

Zone	Definition
Zone 0	This zone is the interior of the basin of the swimming pool or fountain including any recesses in its walls or floors, basins for foot cleaning and water jets or waterfalls and the space below them.
Zone 1	This zone is limited by: • zone 0 • a vertical plane 2 m from the rim of the basin • the floor, or the surface expected to be occupied by persons • the horizontal plane 2.5 m above the floor or the surface expected to be occupied by persons. Where the swimming pool or fountain contains diving boards, spring boards, starting blocks, chutes or other components expected to be occupied by persons, zone 1 comprises the zone limited by: • a vertical plane situated 1.5 m from the periphery of the diving boards, spring boards, starting blocks, chutes and other components such as accessible sculptures, viewing bays and decorative basins • the horizontal plane 2.5 m above the highest surface expected to be occupied by persons.
Zone 2	This zone is limited by: • the vertical plane external to zone 1 and a parallel plane 1.5 m from the former • the floor or surface expected to be occupied by persons • the horizontal plane 2.5 m above the floor or the surface expected to be occupied by persons.

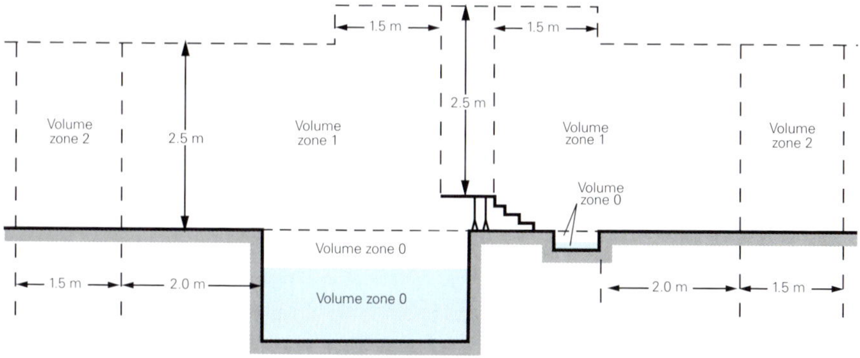

Fig 11.7 Zone requirements for swimming pools and paddling pools

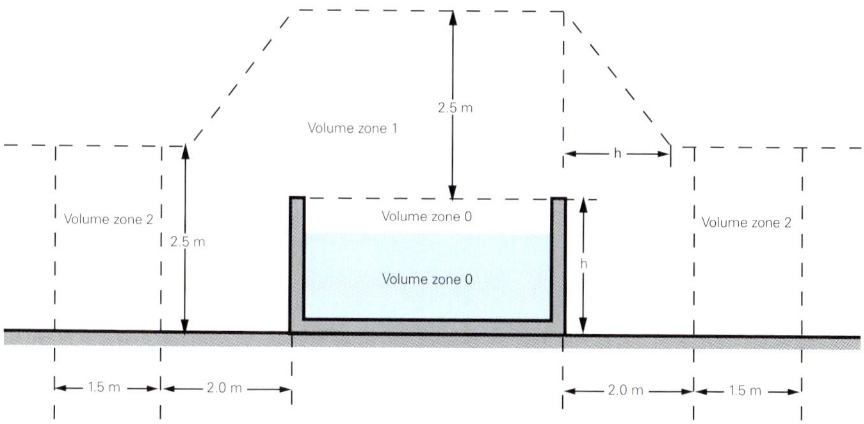

Fig 11.8 Zone dimensions for basin above ground

11.3.2 Protection against electric shock

BS 7671 does **not** allow basic protection to be provided by obstacles or protection by placing out of reach anywhere falling within the scope of Section 702 (Regulation 702.410.3.5 refers). Further, fault protection by non-conducting location or by earth-free local equipotential bonding is similarly prohibited (Regulation 702.410.3.6 refers).

Zones 0 and 1 of a swimming pool

In zone 0, and with the exception of fountains, only protection by SELV at a nominal voltage not exceeding 12 V AC rms or 30 V ripple-free DC is permitted.

In zone 1, and with the exception of fountains, only protection by SELV at a nominal voltage not exceeding 25 V AC rms or 60 V ripple-free DC is permitted.

In both cases, the source for SELV must be installed outside zones 0, 1 and 2.

Any equipment for use in the interior of basins that is only intended to be in operation when people are not inside zone 0 should be supplied by a circuit protected by:

- SELV, having the source for SELV installed outside zones 0, 1 and 2, unless the supply circuit is protected by a 30 mA RCD in which case the source for SELV may be placed in zone 2, or
- automatic disconnection of the supply using a 30 mA RCD, or
- electrical separation supplying only one item of current-using equipment, the source for which installed outside zones 0, 1 and 2, unless the supply circuit is protected by a 30 mA RCD in which case the source for electrical separation may be placed in zone 2.

A warning notice should be located at the socket-outlet used to provide the supply to such equipment and the associated control device to warn the user that this equipment should be used only when the swimming pool is not occupied by persons (see Regulation 702.410.3.4.1 refers).

Zone 2 of a swimming pool

One or more of the following protective measures should be employed in zone 2:

- SELV, having the source for SELV being installed outside zones 0, 1 and 2, unless the supply circuit is protected by a 30 mA RCD in which case the source for SELV may be placed in zone 2, or
- automatic disconnection of supply using a 30 mA RCD, or

- electrical separation supplying only one item of current-using equipment, the source for which installed outside zones 0, 1 and 2, unless the supply circuit is protected by a 30 mA RCD, in which case the source for electrical separation may be placed in zone 2

Note: Where the swimming pool supply forms part of a TN-C-S system employing Protective Multiple Earth (PME) a note to Regulation 702.410.3.4.3 recommends that an earth mat or earth electrode of suitably low resistance (in the region of 20 Ω or less) is installed and connected to the protective equipotential bonding for the swimming pool.

SELV

Where extra-low voltage provided by SELV is used as the protective measure, basic protection must be provided by basic insulation or barriers or enclosures affording a degree of protection of at least IPXXB or IP2X whatever the nominal voltage (Regulation 702.414.4.5 refers).

11.3.3 Supplementary equipotential bonding

If there are any exposed-conductive-parts within zones 0, 1, or 2 (except exposed-conductive-parts of a SELV circuit), local supplementary bonding is required in zones 0, 1 and 2 to connect together the following:

- all extraneous-conductive-parts (which may include metal handrails, exposed structural steelwork and metal pipes etc), and
- the protective conductors of all exposed-conductive-parts within these zones (Regulation 702.415.2 refers)

The connection with the protective conductor may be made in an enclosure for an accessory or in a local distribution board.

Attention should be paid to the protection of supplementary bonding conductors against mechanical damage and to the avoidance of corrosion, particularly at connections.

Connections should be accessible to permit inspection testing and maintenance (see Section 10.7.6 of this Guide). If a metallic grid is installed in a solid floor in zones 0, 1 or 2 (Fig 11.9), it must be connected to the local supplementary bonding. There is no absolute requirement in *BS 7671* to provide a grid, but the need for such a provision should be considered at the design stage, particularly where a solid floor may be an extraneous-conductive-part (for instance, where the floor is not electrically well insulated from Earth potential).

Two or more bonding connections should be made to the grid, preferably at diagonally opposite corners, and it should be ensured that all parts of the grid are connected together such that reliable electrical continuity is provided.

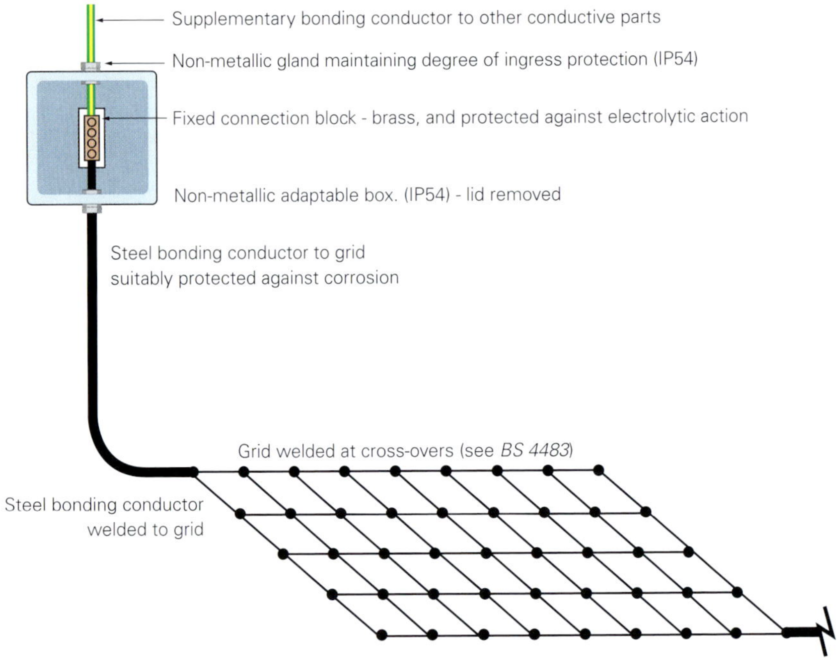

Fig 11.9 Example of one of the connections to a metallic grid

11.3.4 External influences

Each component part of the electrical installation is to be selected and erected to comply with the requirements of Section 522 (Selection and erection in relation to external influences), taking into account all the conditions existing or likely to exist at the point of installation.

Particular areas of concern include:

- the presence of water (Regulation Group 522.3 refers), and
- the presence of corrosive or polluting substances, such as chlorine (Regulation Group 522.5 refers).

11.3.5 Degrees of protection of enclosures against water ingress

The degree of protection of enclosures against ingress of water should be appropriate to the zonal location of the equipment (Regulation 702.512.2 refers).

The minimum permitted degree of protection related to the zone is shown in Table 11.6.

Table 11.6 Minimum degree of protection

Zone	Minimum degree of protection
0	IPX8
1	IPX4 IPX5 where water jets are likely to be used for cleaning
2	IPX2 for indoor pools IPX4 for outdoor locations IPX5 where water jets are likely to be used for cleaning

11.3.6 Wiring systems

Regulation 702.522.21 states that any metallic sheath or metallic covering of a wiring system situated in zones 0, 1 or 2 must be connected to the supplementary equipotential bonding. The regulation further recommends that cables should preferably be installed in conduits made of insulating material.

Within zones 0 and 1 of a swimming pool, only wiring systems necessary to supply equipment situated in these zones should be installed (Regulation 702.522.22 refers).

Junction boxes should not be installed within zones 0 or 1, unless wiring therein is limited to that of SELV circuits, in which case junction boxes may be installed in zone 1 (see 702.522.24).

11.3.7 Switchgear and controlgear

The requirements for switchgear and controlgear in a swimming pool are summarised in Table 11.7.

Table 11.7 Switchgear and control gear

Zone	Requirement	Regulation
0	Switchgear or controlgear not to be installed Socket-outlets not to be installed	702.53
1	Switchgear or controlgear not to be installed Socket-outlets not to be installed For a swimming pool where it is not possible to locate a socket-outlet or switch outside zone 1, a socket-outlet or switch, preferably having a non conductive cover or coverplate, is permitted in zone 1 if it is installed outside (1.25 m) from the border of zone 0, and placed at least 0.3 m above the floor, and it must be protected by: • SELV, at a nominal voltage not exceeding 25 V AC rms or 60 V ripple-free DC, the source for SELV being installed outside zones 0 and 1, or • Automatic disconnection of supply, using a 30 mA RCD, or • Electrical separation for a supply (having a 30 mA RCD) to only one item of current-using equipment or socket-outlet, the source for electrical separation being installed outside zones 0.	702.53
2	A socket-outlet or a switch is permitted only if the supply circuit is protected by one of the following protective measures: • SELV, the source of SELV being installed outside zones 0, 1 and 2. However, it is permitted to install the source of SELV in zone 2 if its supply circuit is protected by a 30 mA RCD, or • Automatic disconnection of supply, using a 30 mA RCD, or • Electrical separation, the source for electrical separation supplying only one item of current-using equipment, or one socket-outlet, installed outside zones 0, 1 and 2. However, it is permitted to install the source in zone 2 if its supply circuit is protected by a 30 mA RCD. The general requirements of *BS 7671* apply.	702.53

11.3.8 Current-using equipment of swimming pools

The requirements for switchgear and controlgear in a swimming pool are summarised in Table 11.8.

Table 11.8 Current-using equipment

Zone	Requirement	Regulation
0	It is only permitted to install fixed current-using equipment specifically designed for use in a swimming pool	702.55.1
1	It is only permitted to install fixed current-using equipment specifically designed for use in a swimming pool	702.55.1
2 Outside of the zone	The general requirements of *BS 7671* apply.	

Any equipment that is only intended to be in use when people are not inside zone 0 may be employed in all zones provided that it is supplied by a circuit protected according to Regulation 702.410.3.4.

Electric heating may be embedded in the floor, provided that one or more of the following requirements of Regulation 702.55.1 is met:

- protection by SELV is provided, the source of SELV being installed outside zones 0, 1 and 2. (the source of SELV may be installed in zone 2 if its supply circuit is protected by a 30 mA RCD), or
- the heating element incorporates an earthed metallic sheath that is connected to the supplementary equipotential bonding specified in Regulation 702.415.2 and its supply circuit is additionally protected by a 30 mA RCD, or
- it is covered by an embedded earthed metallic grid that is connected to the supplementary equipotential bonding specified in Regulation 702.415.2 and its supply circuit is additionally protected by a 30 mA RCD.

11.3.9 Underwater luminaires for swimming pools

Any luminaire for use submerged in the water, or in contact with the water, must be fixed and must comply with *BS EN 60598-2-18: 1994 Luminaires. Particular requirements. Luminaires for swimming pools and similar applications (as amended).*

Any underwater lighting located behind watertight portholes, and arranged to be serviced from behind, should comply with the appropriate part of *BS EN 60598* and be installed such that no intentional or unintentional conductive connection can occur between any exposed-conductive-part of the underwater luminaires and any conductive parts of the portholes (Regulation 702.55.2 refers).

11.3.10 Special requirements relating to the installation of electrical equipment in zone 1 of swimming pools

Fixed equipment designed for use in swimming pools and other basins such as filtration systems, jet stream pumps and the like supplied at low voltage may be installed in zone 1, if all the following requirements given in Regulation 702.55.4 are met:

a) the equipment is located inside an insulating enclosure providing at least Class II or equivalent insulation and providing protection against mechanical impact of medium severity (AG2) irrespective of the classification of the equipment, and

b) the equipment is only accessible via a hatch (or a door) by means of a key or a tool. The act of opening of the hatch (or door) should disconnect all live conductors. Further, the supply cable and the main disconnecting means must be installed such that protection of Class II or equivalent insulation is provided, and

c) the supply circuit of the equipment must be protected by:

   - SELV at a nominal voltage not exceeding 25 V AC rms or 60 V ripple-free DC, the source of SELV being installed outside zones 0, 1 and 2, or
   - a 30 mA RCD, or
   - electrical separation, the source for electrical separation supplying a single fixed item of equipment and being installed outside of zones 0, 1 and 2.

In the case of swimming pools where there is no zone 2, lighting equipment supplied by other than a SELV source at 12 V AC rms. or 30 V ripple-free DC may be installed in zone 1 on a wall or on a ceiling provided that the following requirements are fulfilled:

- the circuit is protected by automatic disconnection of the supply and additional protection is provided by a 30 mA RCD, and
- the height from the floor is at least 2 m above the lower limit of zone 1.

In addition, every luminaire must have an enclosure providing Class II or equivalent insulation and providing protection against mechanical impact of medium severity.

11.3.11 Accessibility

As required by Regulation 513.1, all electrical equipment must be arranged so as to facilitate its operation and maintenance and to permit access to each connection. For example, luminaires should not be mounted directly over the water of a swimming pool unless proper arrangements are in place to gain adequate and safe access to them without undue difficulty.

The facility for access must not be significantly impaired by mounting equipment in an enclosure.

11.3.12 Protective Multiple Earthing (PME)

BS 7671 does not preclude the use of the earthing facility of a PME supply (that is the PME earthing terminal) as the means of earthing for the electrical installation in a location containing a swimming pool or paddling pool.

However, electricity distributors are sometimes unwilling to allow such use, or may attach conditions to it. This is possibly because of complaints, received in the past, of perceived electric shock by persons in such locations, when touching exposed-conductive-parts or extraneous-conductive-parts connected to the PME earthing terminal whilst in contact with an uninsulated solid floor. Such sensation of electric shock is caused by the potential difference that can exist between the PME earthing terminal and 'true' Earth. The potential difference is due to the voltage drop in the PEN conductor, caused by load current returning to the source of supply.

In order to avoid the risk of perceived electric shock caused by the above, it may be preferable to separate the exposed-conductive-parts and extraneous-conductive-parts in the special location from the PME earthing terminal. Often, the installation in the special location would then be arranged to form part of a TT system and would have to meet all the associated requirements of *BS 7671*, including being connected to Earth by a suitable installation earth electrode.

However, avoiding the risk of perceived electric shock by separation of the conductive parts from the PME earthing terminal, as described above, is sometimes impracticable. This is the case, for example, where the special location occupies part of a building, the electrical installation in another part of which uses the PME earthing terminal as its means of earthing. In these circumstances, exposed-conductive-parts and / or extraneous-conductive-parts in the special location might be in metal-to-metal contact with metalwork connected to the PME earthing terminal, meaning that any desired separation is not achieved.

Alternatively, exposed-conductive-parts or extraneous-conductive-parts in the special location might be simultaneously accessible with metalwork connected to the PME earthing terminal, and would therefore, be required to be electrically connected to that metalwork, due to the requirements of Regulation 411.3.1.2 (protective equipotential bonding).

Where it is necessary to use the PME earthing terminal as the means of earthing for the installation in the special location (which should only be done with the distributor's express consent), metallic floor grids connected to the local supplementary bonding should be provided in all uninsulated solid floors. In addition, the distributor may have further requirements, such as the provision of an earth electrode connected to the main earthing terminal of the installation in the special location.

11.4 Saunas

A room or cabin containing a sauna heater is considered by *BS 7671* to be a location of increased electric shock risk, as large areas of the human body are exposed and are wet due to perspiration and occasional high humidity.

Regulation 703.1 states that the particular requirements of Section 703 of *BS 7671* apply to sauna installations erected on site, whether cabins or within a room.

In the latter case the whole room is considered to be the sauna and the requirements of Section 703 need to be applied as appropriate.

The requirements of Section 703 do not apply to prefabricated sauna cabins complying with a relevant equipment standard.

Electrical installations for rooms and cabins containing sauna heaters are also subject to the general requirements of *BS 7671* (supplemented or modified where appropriate by Section 703).

Where rooms and cabins containing sauna heaters form part of another special location, such as a swimming pool, it will be necessary to consider the requirements of *BS 7671* for both locations when designing the installation.

11.4.1 Temperature zones

The interior of a hot air sauna is classified into three zones (1, 2 and 3) for the purposes of protection against electric shock and selecting and erecting suitable equipment. The zones illustrated in Figure 703 of *BS 7671*, are represented in Fig 11.10.

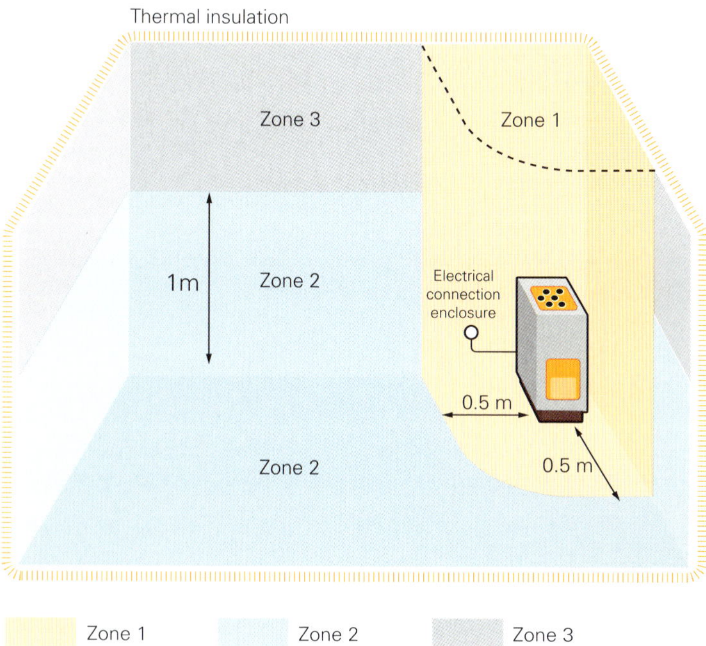

Fig 11.10 Zones of a room and cabin containing a sauna heater

11.4.2 Protection against electric shock

BS 7671 does not allow basic protection to be provided by obstacles or protection by placing out of reach (Regulation 703.410.3.5 refers). Further, fault protection by non-conducting location or by earth-free local equipotential bonding is similarly prohibited (Regulation 703.410.3.6 refers).

11.4.3 Additional protection by RCDs

In general, additional protection must be provided for all circuits of the sauna, by the use of one or more 30 mA RCDs. However, it is not necessary to provide RCD protection for the sauna heater unless recommended by the equipment manufacturer (Regulation 703.411.3.3 refers).

11.4.4 SELV or PELV

Where SELV or PELV is used, regardless of the nominal voltage, basic protection must be provided by basic insulation, or barriers or enclosures affording a degree of protection of at least IPXXB or IP2X (Regulation 703.414.4.5 refers).

11.4.5 Selection and erection of equipment

The general philosophy of Section 703 is to keep all electrical equipment out of the sauna except for equipment which is absolutely necessary for the operation and use of the sauna. Sauna heating appliances must comply with *BS EN 60335-2-53: 2011 – Household and similar electrical appliances. Safety.* Particular requirements for sauna heating appliances and infrared cabins and should be installed in accordance with the manufacturer's instructions.

All equipment in the room or cabin containing the sauna heater must have a degree of protection of at least IPX4 and if cleaning by the use of water jets may to be expected, electrical equipment must have a degree of protection of at least IPX5.

- in zone 1, only the sauna heater and equipment belonging to the sauna heater may be installed
- in zone 2, there are no special requirements relating to the heat resistance of equipment
- in zone 3,
  - » the equipment should be able to withstand a minimum temperature of 125 °C, and
  - » the insulation and sheaths of cables should be able withstand a minimum temperature of 170 °C.

11.4.6 Wiring systems

The wiring system preferably should be installed outside the zones of the location, that is, on the cold side of the thermal insulation. A wiring system installed on the warm side of thermal insulation in zones 1 or 3 must be heat-resisting. Where a wiring system has metallic sheathing or where metallic conduit has been employed, they should not be accessible in normal use (see Regulation 703.52 refers).

11.4.7 Isolation, switching, control and accessories

Switchgear and controlgear forming part of the sauna heater equipment or of other fixed equipment installed in zone 2, should be installed within the sauna room or cabin in accordance with the manufacturer's instructions. Other switchgear and controlgear, provided for control of local lighting, should be installed outside the sauna room or cabin.

Socket-outlets should not be installed within the location (Regulation 703.537.5 refers).

11.5 Lighting installations

In this Guide, details of the requirements for outdoor lighting installations and extra-low voltage lighting installations can be found in sections 12.3.5 and 9.5.4 respectively.

11.6 Electric vehicle charging installations

Section 722 of *BS 7671*, contains additional requirements for electric vehicle (EV) charging installations. The requirements of Section 722 supplement or modify the general requirements contained in the parts 1 to 6 of *BS 7671*.

The requirements of Section 722 do not apply to inductive charging of electric vehicles or to installations for charging mobility scooters or similar vehicles of 10 A and less (Regulation 722.1 refers).

11.6.1 Electric vehicle charging equipment suitable for use in domestic premises

This section of the Guide will focus on the installation of single-phase charging units, of the type designed for use in domestic premises, which are typically rated at 16 A.

The batteries in an electric vehicle are charged from the electricity supply by the use of a flexible cable which is connected to the electric vehicle by means of a vehicle coupler (part of the charging cable) and a vehicle connecter (mounted on the vehicle) inlet.

Regulation 722.55.101.0.201.1 requires that the vehicle coupler and vehicle connecter meet the requirements of *BS EN 62196 - Plugs, socket-outlets, vehicle connectors and vehicle inlets*. There are three designs of connector known as type 1, type 2 and type 3, which meet the requirements of *BS EN 62196*. However, type 2 is the most common connector choice in Europe and an example is shown in Fig 11.11.

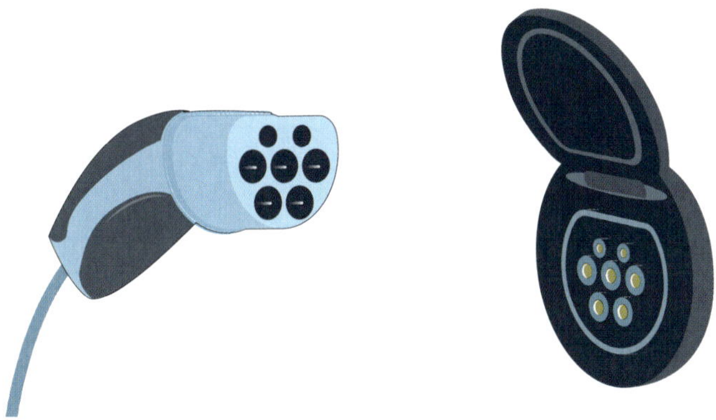

Fig 11.11 Type 2 electric vehicle plug and socket

The charging cable may be connected to the electrical supply by either a *BS 1363-2* or a *BS EN 60309-2* plug and socket-outlet (see Fig 11.12), supplied by a dedicated final circuit or by connection to an electric vehicle charging unit.

Fig 11.13 shows that where an electric vehicle charging unit is used, the charging cable can be either permanently connected to the electric vehicle charging unit (tethered) or be connected to the charging unit via a *BS EN 62196* socket-outlet mounted on the charging unit (Regulation 722.55.101.0.201.1 refers). It should be noted that portable socket-outlets are not permitted. Also, a socket-outlet or vehicle connector shall supply only one electric vehicle (Regulation 722.55.101.3 refers).

Fig 11.12 *BS 1363-2 and *BS EN 60309-2*** socket-outlets**

*Where a 13A *BS 1363-2* socket-outlet is being used, Regulation 722.55.101.0.201.1 (i) requires that unless there is no possibility of confusion for its intended use, it shall be suitably marked with a label stating: 'suitable for electric vehicle charging'.

**Where a *BS EN 60309-2* socket-outlet is being used, to prevent contact with live parts of the socket contacts, Regulation 722.55.101.0.201.1 (ii) requires the socket-outlet to incorporate an interlock that complies with one of the two sets of requirements, described in the regulation.

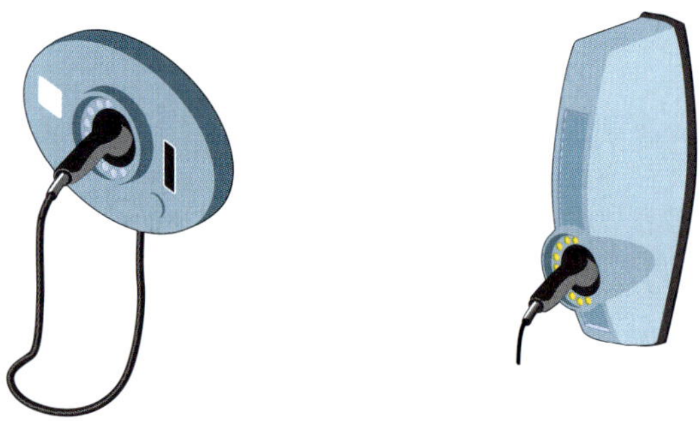

(a) EV charging unit with a tethered charging cable

(b) EV charging unit incorporating a socket-outlet

Fig 11.13 Alternative arrangements for EV charging units

Recent developments in EV charging systems have resulted in several new systems for plugs, socket-outlets and connectors being introduced. If an EV charging system including connectors not recognised by *BS 7671* is to be installed then a note should be attached to the Domestic Electrical Installation Certificate or Electrical Installation Certificate stating 'The EV charging system includes connectors not recognised by *BS 7671*: this does not imply that these connectors are unsafe.

BS 7671 defines an electric vehicle charging point as the point where the electric vehicle is connected to the fixed electrical installation. Where the charging cable belongs to the vehicle, the charging point is a socket-outlet and where the charging cable is permanently connected to the charging equipment, the charging point is the connector of the charging cable.

As electric vehicle charging technology has developed, various standard methods of charging known as modes of charging have been developed and Part 2 of *BS 7671* defines four modes of charging (mode 1 to mode 4). However, the majority of new charging equipment used in a domestic situation is likely to employ either mode 2 or mode 3 charging.

Mode 2 charging: uses a charging cable incorporating an in-cable control box that includes an RCD to connect the EV to a socket-outlet conforming to *BS 1363-2* or *BS EN 60309-2* (see Fig 11.14). The control box also includes a system for detecting safety-specific issues, including that a vehicle is plugged in and that the protective earth is connected.

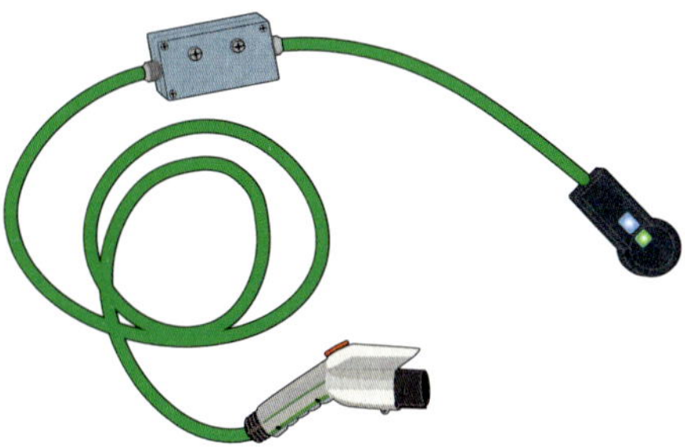

Fig 11.14 – A mode 2 charging cable

Although some vehicle manufacturers may offer a mode 2 charging cable at an entry level, most car manufacturers would recommend dedicated mode 3 charging systems. However, the ability to plug into any available *BS 1363-2 13* A socket-outlet means that some owners of electric vehicles use mode 2 cables as standby equipment, that is for use when the battery becomes depleted and there is no dedicated EV charging point within range.

Mode 3 charging, compared to mode 2 provides for faster charging. Although on some systems one end of the charging cable may be permanently tethered to the charging point or vehicle, others use plugs and socket-outlets complying with *BS EN 62196*. The mode 3 charging equipment may well include an RCD and the electronics required to monitor safe operation. To enable the equipment to communicate with the vehicle, the cable and its connectors are provided with additional cable cores.

Vehicles produced by different manufacturers may well use different connector types, and before installing electric vehicle charging equipment, as advised by a note to Regulation 722.55.101.0.201.1, the installer should consult the vehicle manufacturers' instructions.

11.6.2 The supply to electric vehicle charging equipment in domestic premises

Circuit design and loading allowances

Regulation 722.311 requires that a final circuit provided for the connection to electric vehicles must be dedicated to that purpose alone. Where a final circuit supplies one or more charging points, the regulation states that an allowance for diversity is not permitted.

Protective measures against electric shock

Regulation Group 722.4 identifies a number of protective measures that shall not be used. Therefore, it is expected that the protective measure for EV charging equipment installed in domestic premises shall be Automatic Disconnection of Supply (as described in Section 411).

Where Protective Multiple Earthing (PME) is used

As shown in Fig 11.15, where the protective measure is automatic disconnection of supply and the means of earthing is PME (TN-C-S system), in the event of the combined protective and neutral (PEN) conductor supplying the installation developing an open-circuit fault, the earthed metalwork of the installation and any equipment connected to it (including any electric vehicle and its charging equipment) can rise to a dangerous voltage. This dangerous voltage can exist for long periods of time and will pose a danger to any person touching the metalwork whilst in contact directly with the general mass of Earth. In this situation, as the shock current flows in both the line and neutral conductors, there is no imbalance in the currents and an RCD will not interrupt the current.

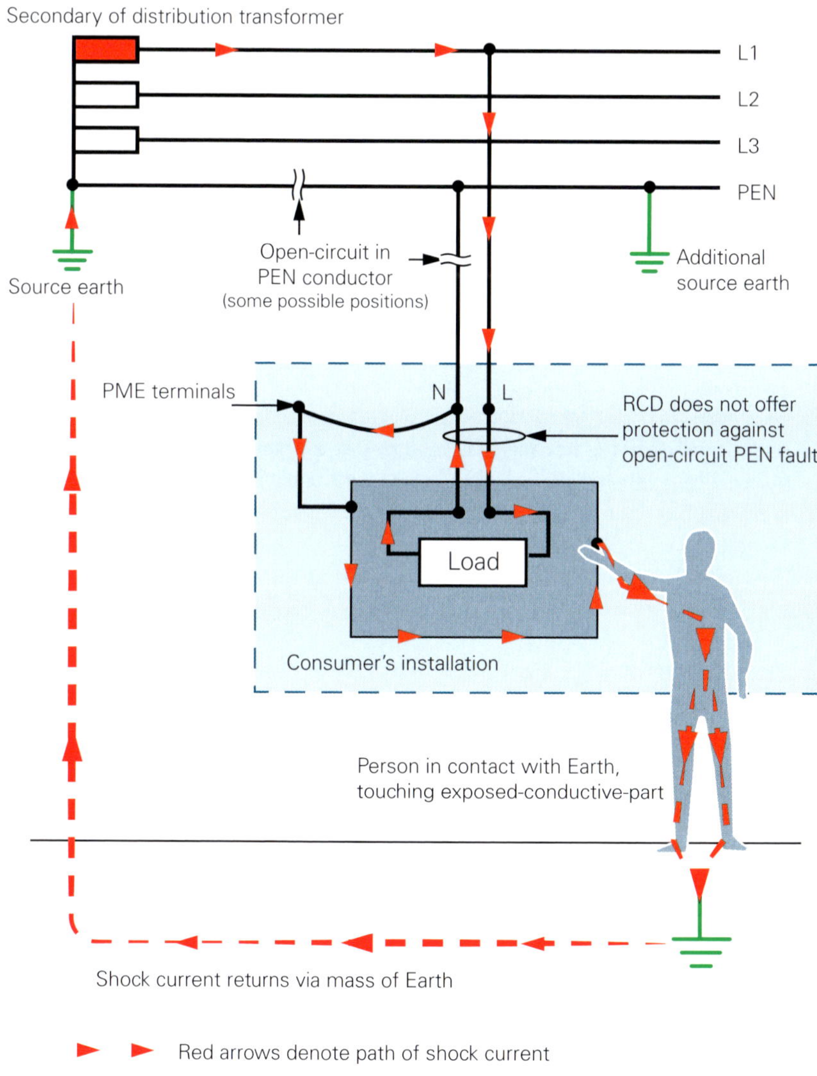

Fig 11.15 Electric shock risk due to an open-circuit fault in the supply PEN conductor

Regulation 722.411.4.1, states that where the electric vehicle charging point is located outdoors, or might reasonably be expected to be used to charge a vehicle located outdoors and the protective measure is automatic disconnection of supply, a PME earthing facility is only permitted where one of the following conditions apply.

(i) The charging point forms part of a three-phase installation that also supplies loads other than for electric vehicle charging and because of characteristics of the load and the supply in the event of an open circuit fault in the PEN conductor of the supply to the installation, the voltage between the main earthing terminal and Earth does not exceed 70 V rms.

Item A722.2 of Annex A722 provides further guidance.

(ii) The main earthing terminal of the installation is connected to an installation earth electrode by a protective conductor complying with Regulation 544.1.1. The resistance of the earth electrode to Earth shall be such that the maximum voltage between the main earthing terminal of the installation and Earth in the event of an open-circuit fault in the PEN conductor of the low voltage network supplying the installation does not exceed 70 V rms.

Item A722.3 of Annex A722 provides further guidance on determining the maximum resistance required for the earth electrode.

(iii) Protection against electric shock is provided by a device which disconnects the charging point from the live conductors of the supply and from protective earth in accordance with Regulation 543.3.3.101 (ii) within 5 s in the event of the voltage between the circuit protective conductor and Earth exceeding 70 V rms. The device shall not operate if the voltage exceeds 70 V rms for less than 4 s. The device shall provide isolation. Closing or resetting of the device shall be by manual means only. Equivalent functionality could be included within the charging equipment.

RCD protection of charging points

Where automatic disconnection of supply is used as the protective measure, Regulation 722.531.2.101 requires every charging point to be individually protected by an RCD having a rated residual operating current ($I_{\Delta n}$) not exceeding 30 mA. The device must disconnect all live conductors; including the neutral (Regulation 722.531.2.1.1 refers).

The RCD is required to be at least a type A, due to the likelihood of DC being present in vehicle charging current. If it is known that the DC component of residual current will exceed 6 mA, the RCD must be of type B to *BS EN 62423*. Types of RCD are discussed in Section 3.5.3 of this Guide.

It should be noted that the use of a mode 2 charging cable, which includes an RCD, does not meet the requirements of Regulation 722.531.2.101. However, where an EV charging unit (as shown in Fig 11.16) is used, the unit may include a suitable RCD.

Protection by electrical separation

Regulation 722.413.1.2 permits the use of the protective measure electrical separation. This measure is limited to supply of one electric vehicle from one unearthed source. Furthermore, such a circuit must be supplied through a fixed isolating transformer complying with *BS EN 61558-2-4*.

The use of this protective measure to supply an electric vehicle, therefore, does not require the use of an earth electrode.

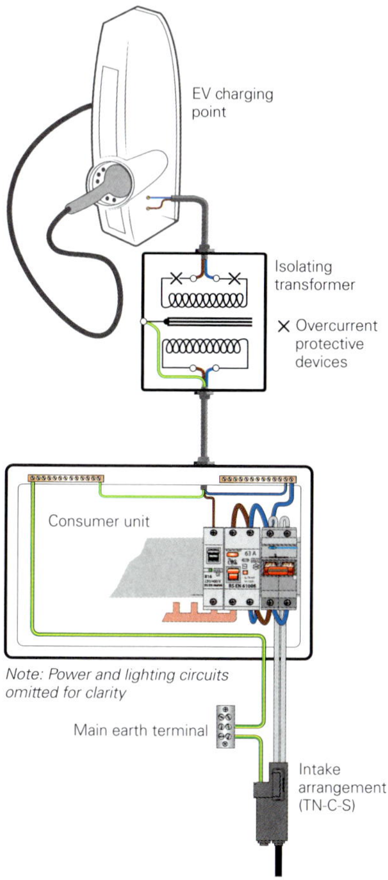

Fig 11.16 Electrical separation arrangement for a vehicle charging point

Protection against external influences

Where an EV charging unit is installed outdoors, the equipment is required to be selected with a degree of protection of at least IPX4.

Miscellaneous requirements

Other notable requirements of Section 722 include:

- Emergency switching devices, where required, must be capable of breaking the full load current of the relevant parts of the installation and disconnect all live conductors (Regulation 722.537.4.101 refers).
- Each socket-outlet shall be installed in a distribution board in accordance with Regulation 722.51 or in its appropriate enclosure, such as a flush or surface mounted socket-outlet box, and mounted in a fixed position (Regulation 722.55.101.0.201.2 refers).
- Tethered vehicle connectors are permitted but not portable socket-outlets (Regulation 722.55.101.0.201.2 refers).
- For charging equipment employing charging modes 3 or 4, unless the socket-outlet or the vehicle connector has been de-energised by switched off from the supply, arrangements must be provided to prevent insertion or removal of plugs (Regulation 722.55.101.4 refers).
- The lowest part of a socket-outlet is to be between 0.5 m and 1.5 m above the ground (Regulation 722.55.101.5 refers).

CHAPTER 12
ELECTRICAL EQUIPMENT OUTSIDE OF THE DWELLING

12.1 Increased risk of electric shock

There is no specific section in Part 7 of *BS 7671* for installations in gardens or other areas external to the premises. However, the dangerous effects of electric shock are greater for persons in contact with the general mass of Earth. This chapter explains the reason for this increase in risk and deals with the related requirements in *BS 7671* intended to address the risk.

Persons within a building are often insulated from the general mass of Earth by electrically insulating floors such as wood, floor coverings such as carpet, and dry footwear. However, persons in outhouses, garages, sheds or similar locations with earthen or stone floors, and particularly persons outdoors (as in Fig 12.1), are likely to be in contact directly with the general mass of Earth.

When a person is standing on, or otherwise touching, the ground, the resistance of the electrical connection between the person's body and the general mass of Earth is significantly reduced. This results in an increased current through the body to Earth, leading to a greater risk of death or serious injury, in the event of either:

- contact of the person with a live part, or
- contact of the person with an exposed-conductive-part, such as the metallic casing of a Class I appliance, that has become live under single fault conditions.

In addition, the wet conditions to which a person may be exposed outdoors mean that shoes and feet, as well as hands and other parts of the body, could be wet, reducing contact resistance. The person may even be barefoot. The result is that the resistance of the path through the body is further reduced, and the current through the body to Earth in the event of failure of basic insulation or under fault conditions is greater, resulting in increased danger.

Clearly, a person may receive an electric shock from mobile electrical equipment (such as a lawnmower, hedge trimmer or power drill) or from the flexible cable supplying such equipment.

Mobile equipment is electrical equipment intended to be moved while in operation or which can easily be moved from one place to another whilst still connected to the supply.

Fig 12.1 Person using electrical equipment whilst standing on the general mass of Earth

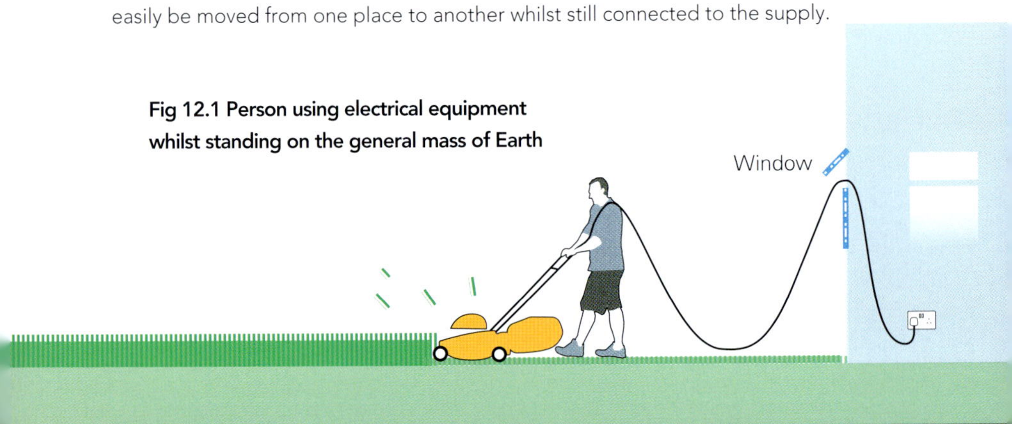

12.2 Additional protection by means of an RCD

Regulation 411.3.3 requires **all** socket-outlets rated at 32 A or less and mobile equipment with a current rating not exceeding 32 A for use outdoors, to be provided with additional protection by means of an RCD having the characteristics specified in Regulation 415.1.1.

Compliance with Regulation 411.3.3 may be achieved by the installation of RCDs such as RCCBs, RCBOs or socket-outlets with integral RCDs (SRCDs), examples of which are illustrated in Figs 12.2 and 12.3.

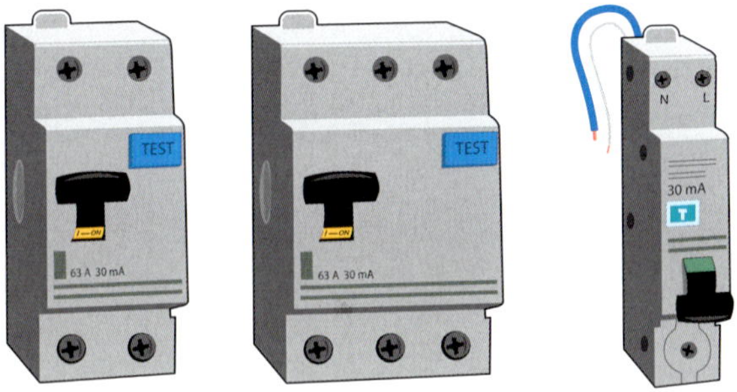

Fig 12.2 Examples of single-phase and three-phase RCCBs, and an RCBO (shown on right)

Fig 12.3 SRCD (Socket-outlet with RCD)

12.3 Electrical equipment

All equipment installed outside of the premises should be designed for use outdoors and be installed in accordance with the manufacturer's instructions. In some cases, additional measures may need to be taken to protect equipment that is not of itself entirely suitable for installation outdoors. For example, equipment may need to be so placed that it is not directly exposed to the effects of the weather or be placed within an appropriately weatherproof enclosure (see also Fig 12.3 which shows a SRCD in such an enclosure).

The use of Class II equipment is conducive to electrical safety in such locations and is to be encouraged.

12.3.1 Buried cables

Where cables are buried in the ground, they should be installed in such a manner that they will not suffer damage during the course of any reasonably foreseeable disturbance of the ground above (such as digging over a flower bed or spiking a lawn).

This may affect the choice of cable routes adopted. It will also require the cables to be buried sufficiently deep into the ground. In general, a minimum depth of 600 mm should be acceptable in a domestic garden.

The position of all buried cables should be marked by cable covers or marker tape (see Fig 12.4).

Non-armoured buried cables should be installed in a conduit or duct that provides protection equivalent to armour against mechanical damage (Regulation 522.8.10 refers).

Steel wire armoured (SWA) cables and cables having an earthed metal sheath suitable for use as a protective conductor having an outer plastic covering may be buried directly in the ground without further protection.

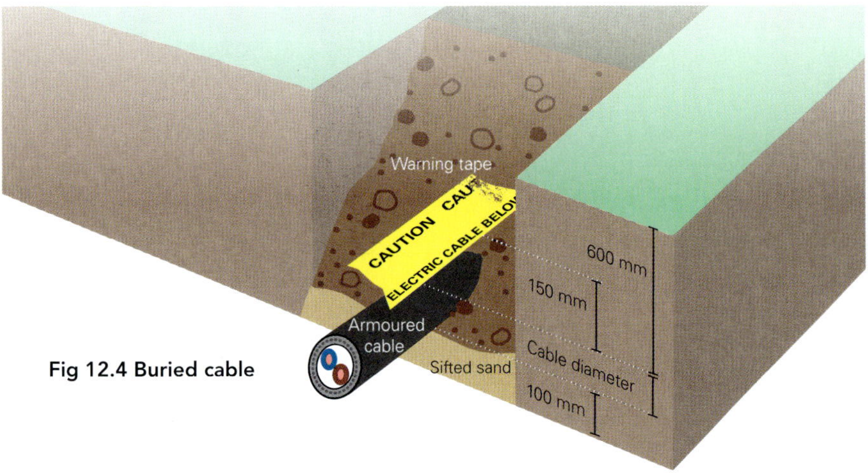

Fig 12.4 Buried cable

12.3.2 Surface-installed cables

Surface-installed cables should be fixed to permanent structures only and should be supported at sufficient height above the ground to avoid accidental contact with footwear and garden tools, which could damage the cable.

Timber fencing, as shown in Fig 12.5, is not considered to be suitable for supporting cables for reasons such as:

- wood preservatives could attack the cable sheath,
- timber fencing is unlikely to be sufficiently rigid,
- the life of timber fencing is likely to be less than that of the cable,
- damage likely to occur to the fence due to the high winds (that may present a risk of electric shock).

Surface-installed cables should not be exposed to prolonged sunlight unless they are of a suitable type. Black thermoplastic (PVC) cable sheaths contain carbon black which resists the effects of sunlight. White or grey cable sheaths are likely to deteriorate with prolonged exposure to the sun and should always be shielded to avoid this happening. See Chapter 10 for more details.

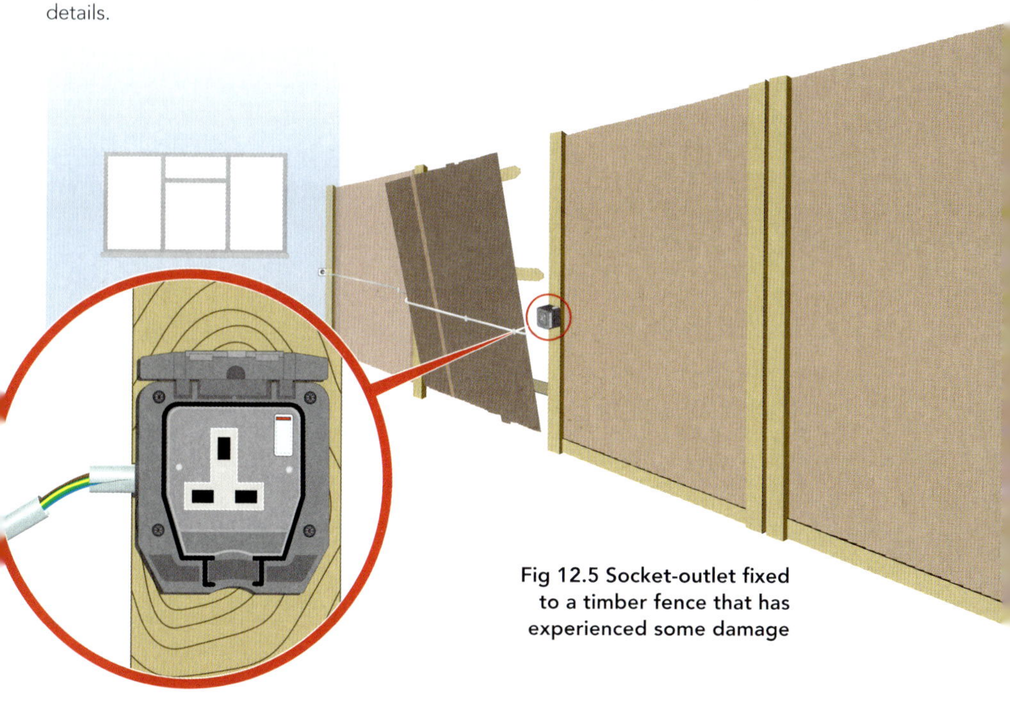

Fig 12.5 Socket-outlet fixed to a timber fence that has experienced some damage

12.3.3 Armoured cable terminations

Where SWA cables are terminated in a location where they are exposed to the elements, weatherproof glands should be used. In such cases, and where practicable, cables should not be terminated into the top surfaces of enclosures.

A shroud should be installed when terminating an SWA cable, but should not be so tight that it impairs future inspection of the gland.

12.3.4 Socket-outlets

Provided they are designed for outdoor use, 13 A socket-outlets to *BS 1363* may be installed in a garden as a convenient way of connecting luminaires and other equipment. 13 A socket-outlets installed in a garden must either incorporate a 30 mA RCD or be supplied from a circuit provided with additional protection by a 30 mA RCD (Regulation 411.3.3 refers).

Whilst for household installations, only shuttered socket-outlets should be used (see Regulation 553.1.201), industrial socket-outlets including plugs, connectors and appliance inlets complying with *BS EN 60309-2* may be used in locations outside of the dwelling where appropriate.

12.3.5 Lighting

Section 714 contains additional requirements for outdoor lighting installations. However, as described by Regulation 714.1, Section 714 does not apply to luminaires fixed to the outside of a building and supplied directly from the internal wiring of that building or to temporary festoon lighting such as Christmas lights. Section 714 does apply to circuits and electrical equipment used for general external lighting and to situations where lighting used to illuminate features in the garden or to provide floodlighting.

Many of the regulations contained in Section 714 relate to street furniture, street lighting, illuminated advertising panels and the like and do not apply to the outdoor lighting of domestic premises. The main requirements of Section 714 that are applicable to the outdoor lighting installations of domestic premises are as follows.

1. Where the protective measure automatic disconnection of supply is used, all live parts of electrical equipment shall be protected by insulation or by barriers or enclosures providing basic protection (Regulation 714.411.201 refers).
2. The protective measures non-conducting location and earth-free equipotential bonding shall not be used. (Regulation 714.410.3.6 refers).

3. By necessity enclosures forming part of the outdoor lighting installations must be designed to prevent contact with live parts and where the protective measure is Automatic Disconnection of the Supply, Regulation group 714.411 requires that unless contained in an enclosure that is located where it is only accessible by skilled or instructed persons live parts shall only be accessible by use of a key or a tool. Furthermore Regulation 714.411.2.201 requires that where a luminaire is located at less than 2.8 m above the ground, the lamp should be accessible only after removing a barrier or opening an enclosure requiring the use of a tool.
4. Only lighting equipment designed for external use should be installed outdoors. The requirements applicable to the lighting equipment being used outdoors is summarised in Table 12.1.

Table 12.1 Summary of external influence requirements of Regulation 714.512.2 applicable to outdoor lighting installations

	Ambient temperature	Relative humidity	Ingress protection		Degree of protection (IP rating) required for enclosure
			Water	Foreign bodies	
Classification (See also Appendix 5 of *BS 7671*)	AA2 to AA4	AB2 to AB4	AD3	AE2	IP33
Description	-40 °C to +40 °C	5 % to 100 %	Resistant to sprays	Small objects (2.5 mm)	Protected against: access to hazardous parts with a tool or spray at an angle of up to 60° of vertical

5. Regulation 714.537.2.1.1 requires that, except as described in Regulation 461.2, every circuit shall be capable of being isolated from each of the live supply conductors. Regulation 461.2 recognises that in TN-S or TN-C-S systems, the neutral conductor is reliably connected to earth, and need not be isolated. Consequently outdoor lighting supplied by a single-phase TN-S or TN-C-S system can be isolated by a single-pole device in the live conductor but outdoor lighting supplied by a single-phase TT system should be isolated using a double-pole device.

Regulation 411.3.4 requires that, within domestic premises, additional protection by an RCD with a rated residual operating current not exceeding 30 mA shall be provided for AC final circuits supplying luminaires.

In addition, lighting equipment used outdoors should be securely fixed and protected against accidental damage and attack by rodents. Luminaires installed at low level may require additional mechanical protection. It is advisable that flexible cables used to connect luminaires to the fixed installation should be kept as short as possible.

Festoon lighting, if used, should be adequately supported and securely fixed in accordance with the manufacturer's instructions.

Additional safety measures may be required where luminaires are installed close to a pond.

Submersible luminaires, if used, must always be installed strictly in accordance with the manufacturer's instructions.

Luminaires covered by the following parts of the *BS EN 60598-2* series are suitable for use in gardens:

- *BS EN 60598-2-7: 1997 – Portable luminaires for garden use.*
- *BS EN 60598-2-13: 2006 (2012) – Ground recessed luminaires.*
- *BS EN 60598-2-20: 2015 – Lighting chains.*

12.4 Equipment for use in ponds

Only special equipment and submersible cables designed for use in ponds should be used. Risk of damage to equipment and cables will be minimised by careful positioning and routing of cables.

Cables may need additional protection against accidental damage and attack by rodents.

All equipment located close to a pond should be protected against ingress of water and be resistant to corrosion.

Pond pumps should comply with *BS EN 60335-2-41: 2003 (2010) - Household and similar electrical appliances. Safety. Particular requirements for pumps.*

Other equipment for ponds should comply with *BS EN 60335-2-55: 2003 (2008) - Household and similar electrical appliances. Safety. Particular requirements for electrical appliances for use with aquariums and garden ponds.*

Pond lighting should comply with *BS EN 60598-2-18: 2004 – Luminaires for swimming pools and similar applications.*

All equipment associated with ponds must be installed strictly in accordance with the manufacturer's instructions.

Electrical installations associated with large ponds, which could be used for bathing, should comply with the additional requirements for swimming pools contained in Section 702 of *BS 7671*.

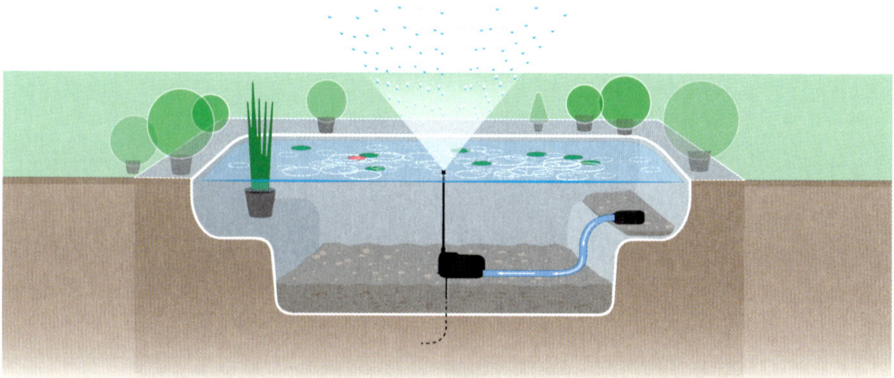

Fig 12.6 Pond with submersible equipment

CHAPTER 13 ADDITIONAL SOURCES OF SUPPLY

13.1 Introduction

In response to increased concerns over the issue of global warming and climate change, and in order to honour commitments made under the Kyoto Protocol, the UK government enacted a number of measures in the Climate Change Act 2008.

Key targets within this Act are:

- to ensure that the net UK carbon account for the year 2050 is at least 80% lower than the 1990 baseline,
- a reduction in emissions of at least 30% against the 1990 baseline by 2020.

At the European level, Directive 2009 / 28/ EC, contains a national overall target for the UK of 20% for the share of energy from renewable sources in gross final consumption of energy in 2020.

As a result of the above commitments, the UK government introduced Feed-in Tariffs (FITs) to provide a financial incentive to encourage uptake of microgeneration.

Electricity-producing microgeneration systems supplement the conventional large power stations with small-scale embedded generators (SSEGs) located in consumers' premises. These microgeneration systems utilise a range of technologies, including photovoltaic (PV), Stirling engine-driven, wind-based and water-based (micro-hydro) systems.

Irrespective of the technology used, microgeneration offers the advantages of zero or extremely low fuel costs and a reduction in the transmission losses that occur in the national transmission of electrical energy.

For microgeneration systems (generally characterised as being rated at up to 16 A per phase) to be connected to the public electricity supply network, the installed equipment must satisfy the requirements of publication *G83 / 2 – Recommendations for the Connection of Type Tested Small-scale Embedded Generators (up to 16 A per phase) in Parallel with Low-Voltage Distribution Systems,* issued by the Energy Networks Association. G83 / 2 is intended to provide manufacturers and installers with guidance on the technical aspects of connecting SSEGs in parallel with the public electricity supply network.

In order to access the funds available through FITs, the installer must, amongst other things, supply evidence to the Distribution Network Operator (DNO), within 30 days of commissioning the microgeneration system that the microgeneration system installation complies with G83 / 2.

It should be noted that payments for SSEGs installed under a FITs scheme can only be made if the work is carried out by an installer registered on the Microgeneration Certification Scheme (MCS) and installed in compliance with the guidance published by MCS appropriate to the type(s) of small-scale generation employed.

In recent years, the planning restrictions applying to the installation of domestic microgeneration systems on the outside of buildings have been relaxed, with the installation of some forms of microgeneration system being classed as permitted development.

However, this relaxation is unlikely to apply in conservation areas, or to a listed building or to a building on a world heritage site. The installation of a microgeneration system in one of these situations will require planning permission and may require listed building consent. Even when the work is classed as permitted development, there may be some planning restrictions in force.

The installation of some forms of microgenerator may be subject to planning law - and to non-electrical aspects of the Building Regulations, particularly in relation to structural considerations.

Therefore, before installing a microgeneration system it would be prudent to check with the local planning office as to what restrictions, if any, apply. Depending upon whether the property is in England, Wales, Scotland or Northern Ireland, different legislation will apply. In Scotland a Building Warrant may be required.

In England and Wales it is a statutory requirement that a microgeneration system complies with the relevant parts of the Building Regulations.

Requirements relating to many of the electrical aspects of installing microgeneration systems are included in *BS 7671* (IET Wiring Regulations 18th Edition), *BS EN 50438: 2013 - Requirements for the connection of micro-generators in parallel with public low-voltage distribution networks* and, depending on the technology employed, requirements of other standards such as *BS EN 62446-1: 2016 Photovoltaic (PV) systems - Requirements for testing, documentation and maintenance. Grid connected systems. Documentation, commissioning tests and inspection.*

13.2 *BS 7671* requirements for generators to operate in parallel

The general rules of *BS 7671* apply to the provision of small-scale generation for domestic premises (Regulation 110.1.1 (i), (vi) & (xviii) refers). For solar photovoltaic (PV) systems the requirements of Section 712 also apply (See Section 13.4 of this Guide).

Regulation 132.2 indicates that a number of characteristics relating to **each** source of supply need to be determined. This requirement is reinforced by Regulation 551.2.2, which requires prospective short-circuit current and prospective earth fault current to be determined for each source of supply.

Regulation 551.2.2 further requires that the short-circuit rating of protective devices within the installation should not be exceeded for any intended methods of operation of the sources.

Wherever an installation is supplied from more than one source, Regulation 537.1.2 states that unless an interlock arrangement is provided (see Section 7.2 of this Guide), a durable warning notice shall be placed in such a position that any person, before gaining access to live parts, will be warned of the need to isolate those parts from the various supplies.

Warning notices are further discussed in Section 8.10 of this Guide.

Section 551 (Low voltage generating sets) contains a number of requirements applicable to installations containing a means of generation that operates in parallel with a supply taken from a public supply network (Regulation 551.1 refers).

Regulation 551.1.1 clarifies that generating sets driven by power sources such as combustion engines, photovoltaic cells and other suitable sources fall within the scope of Section 551.

A generating set should be installed such that:

- it does not impair the safety or proper functioning of other sources of supply (Regulation 551.2.1 refers).
- any RCDs within the installation installed in accordance with Chapter 41 remain effective for every intended combination of sources of supply (Regulation 551.4.2 refers).

It remains necessary to provide protection by automatic disconnection of supply in accordance with Section 411, subject to modification specific to the means of operation of the generator (See Regulation 551.4.3.1 and the associated regulations listed therein). In the case of small-scale embedded generation falling within the scope of this Guide, the requirements of Regulation Group 551.7, which relates to generating sets that may operate in parallel with other sources, should be applied.

These requirements are summarised below:

i) The generator should be installed on the supply side of all protective devices for final circuits within the installation.

ii) Alternatively, except for solar PV systems, the generator may be installed on the load side of the all protective devices for a final circuit of the installation, but only if **all** the following conditions are met:

- the current-carrying capacity (I_z) of the conductors of the final circuit should be equal to or greater than the sum of the rated current of the protective device for the circuit (I_n) and the rated output of the generator (I_g) $I_z \geq I_n + I_g$
- the connection must not be made by plug and socket-outlet

- any RCD providing additional protection of the final circuit must disconnect all live conductors, including the neutral conductor
- there should be no connection between Earth and the live conductors of either the final circuit or the generator
- unless the final circuit protective device disconnects all live conductors, it will be necessary to determine that the sum of the disconnection time for the protective device and the time required for the output voltage of the generating set to fall to 50 V or less does not exceed the appropriate disconnection time required by Regulation Group 411.3.2 for the circuit in question.

When selecting a generator for use in parallel with a public distribution system, consideration should be given factors such as those listed below to avoid adverse effects to the public distribution system and other installations connected to it:

- power factor,
- voltage changes (variations in rms or peak values),
- voltage unbalance (differences between voltage deviations caused by differences in phase currents etc),

  Note: voltage unbalance between phases is permitted for single installations where total output does not exceed 16 A per phase (see clause 5.8 of ENA G83 / 2).
- harmonic distortion,
- issues connected with starting,
- synchronising,
- voltage fluctuation (a series of voltage changes or a cyclical variation).

The use of an automated synchronisation system dealing with frequency, phase and voltage is preferred (Regulation 551.7.3 refers).

Automatic switching must be provided to disconnect the generator from the public distribution system in the event of complete loss, specified degrees of variation of voltage or frequency from declared values (see Regulation 551.7.4). Similarly means should be provided, to prevent the connection of a generator to the public distribution system under such circumstances (Regulation 551.7.5 refers).

Means should be provided in a suitably accessible position, to isolate the generator from the public distribution network (Regulation 551.7.6 refers).

For generators falling within the scope of this Guide (having an output not exceeding 16 A) Regulations 551.7.4, 551.7.5 and 551.7.6 all cross-refer to requirements within *BS EN 50438: 2013 - Requirements for the connection of micro-generators in parallel with public low-voltage distribution networks that should be met.*

13.3 Energy Networks Association G83 / 2

G83/2 - Recommendations for the Connection of Type Tested Small-scale Embedded Generators (Up to 16 A per phase) in Parallel with Low-Voltage Distribution Systems is published by the Energy Networks Association (ENA).

G83/2 contains technical requirements for the design, installation, commissioning and use of such microgeneration systems that must be satisfied before the Distribution Network Operator (DNO) will allow the microgeneration system installation to be connected to the DNO's Network.

A summary of significant content is given below:

Section 5.2 – The installation between the generator and the point at which the supply is delivered to a consumer's installation (the exit point) is to comply with the requirements of *BS 7671*. The generator should be connected directly to an isolator which disconnects all live conductors. A second isolator may be necessary where an inverter is sited remotely, in a loft for example. This section calls for all isolators forming part of the installation to be capable of being locked in the 'off' (isolated) position, which is over and above *BS 7671* requirements.

Section 5.3 – Requires interface protection to be provided to disconnect the generator system from the DNO's distribution system when operating parameters such as loss of mains, voltage variation and frequency variation exceed the settings shown in Table 1 contained in this section of G83 / 2.

The generator protection system must prevent its reconnection to the DNO's distribution network until the voltage and frequency on the DNO's network have remained within the limits stated in Table 1 for a minimum of 20 seconds. This is to prevent a generator being damaged by being reconnected when it is out of synchronism with the rest of the network.

Section 6.2 – Virtually repeats the requirement for a warning notice given in Regulation 514.15 of *BS 7671*, calling for the installer to provide labelling at the fused cut-out, meter position, consumer unit and at all points of isolation between the cut-out and the generator.

Section 6.3 – Recommends that periodic testing of the microgeneration system is carried out at intervals suggested by the manufacturer.

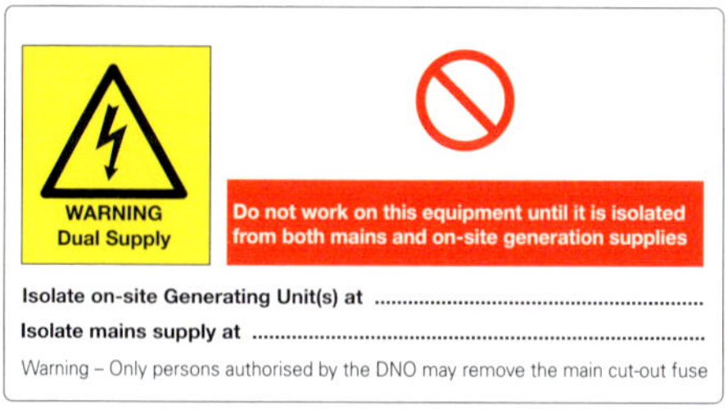

Fig 13.1 Warning notice for installations having more than one source of supply

Section 6.4 – All exposed-conductive-parts shall comply with the earthing requirements of *BS 7671*. There should be no direct connection between the generator current carrying conductors and Earth (see also Regulation 551.7.2 (iii)), except where a generator is connected via an inverter, when it is permissible to connect one pole of the DC side of the inverter to the DNO's earth terminal if there is at least simple separation insulation between the AC and the DC sides (for PV systems see also 712.312.2).

Section 7.2 – Installation should be carried out by persons having recognised and approved qualifications relating to general electrical installation work and the type of microgeneration system being installed.

Previous written agreement should be obtained from the relevant DNO prior to any modification of operating criteria relating to the electrical connection.

A functional check of loss of mains protection should be performed as part of the on-site commissioning tests carried out by the installer.

Section 7.3 – The installer should advise the DNO of the intention to use a microgeneration system in parallel with the Network no later than 28 days (inclusive of the day of commissioning) after commissioning the microgeneration system. The installer should also provide the local DNO with all relevant details on the installation.

Section 7.4 – If a component of a microgeneration system is replaced, it is only necessary to notify the relevant DNO where its operating characteristics or those of the interface protection are different from those in place when the system was originally commissioned.

Section 7.5 – The user of the system (the householder) should notify the DNO if a microgeneration system is decommissioned and will no longer be operated as a source of electrical energy in parallel with a DNO's Network.

13.4 Solar photovoltaic (PV) systems

13.4.1 General

Section 712 of *BS 7671* contains requirements applicable to solar PV power supply systems specifically. These requirements are summarised below.

A solar PV system consists of

- PV modules (panels)
- DC isolator
- inverter
- AC isolator
- generation meter
- AC isolator

And beyond the consumer unit,

- export meter (usually only provided where the installed capacity of the microgenerator exceeds 30 kW)
- import meter

13.4.2 The DC side

A solar array is made up from a number of solar panels, or more correctly PV modules connected together. Each module will contain a number of PV cells that generate electricity when exposed to sunlight.

A number of modules will be connected together in series in order to obtain the required output of voltage.

The PV modules should comply with the requirements of a relevant product standard and should be preferably of Class II construction where the open-circuit voltage of the PV string under standard test conditions ($U_{OC\ STC}$) exceeds 120 V DC (Regulation 712.511.1 refers). Indeed, Class II construction is preferred on the DC side in any case (Regulation 712.412 refers).

The PV modules should be so installed that adequate heat dissipation is ensured for the maximum solar radiation conditions for the location at which they are installed (Regulation 712.512.2.1 refers).

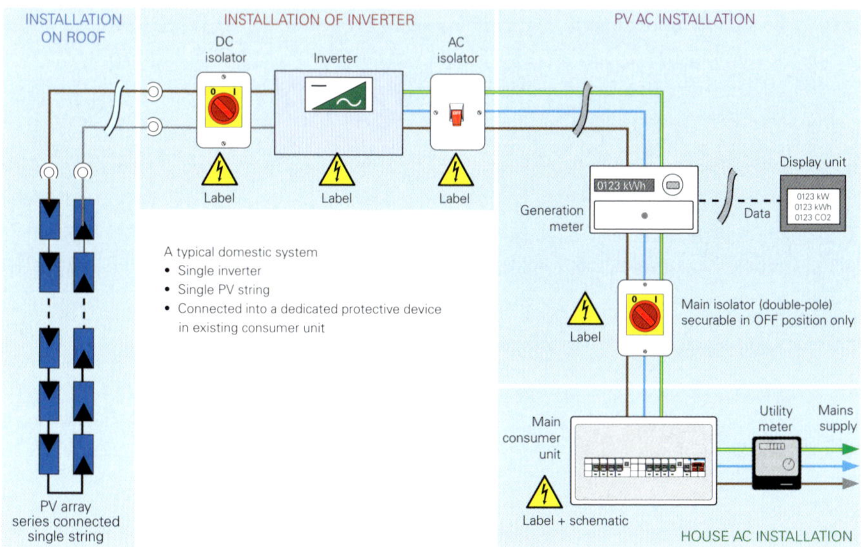

Fig 13.2 Schematic of typical Solar PV microgenerator system

PV string cables, PV array cables and PV DC main cables should be selected and erected such that the risk of earth faults and short-circuits is minimised (Regulation 712.522.8.1 refers). Furthermore, all wiring systems forming part of the solar PV installation should be capable of withstanding the worst conditions of external influence to which they could be expected to be exposed (Regulation 712.522.8.3 refers).

The protective measures of non-conducting location and earth-free local equipotential bonding must not be used on the DC side (Regulation 712.410.3.6 refers).

Overload protection may be omitted for PV string and PV array cables where their continuous current-carrying capacity is greater than or equal to 1.25 times the short-circuit current under standard test conditions ($I_{SC\ STC}$) at any location (Regulation 712.433.1 refers).

No overload protection is required for PV main cables where the continuous current-carrying capacity is greater than or equal to 1.25 times $I_{SC\ STC}$ of the PV generator (Regulation 712.433.2 refers).

13.4.3 Isolation and switching

When considering requirements for isolation and switching, Regulation 712.537.2.2.1 states that the public supply is considered to be the source and the PV installation the load. Any equipment on the DC side, including isolation and switching devices, must be suitable for use with direct current and direct voltage (Regulation 712.512.1.1 refers).

A means of isolation must be provided between both the PV generator and the inverter, and the inverter and the AC side to permit servicing and maintenance (Regulation 712.537.2.1.1 refers). The means of isolation on the DC side must be a switch-disconnector (as discussed in Section 7.4 of this Guide) (Regulation 712.537.2.2.5 refers).

Equipment on the DC side of the PV installation should be considered to be energised even once it has been disconnected from the AC side (Regulation 712.410.3 refers).

Enclosures, junction boxes and the like should comply with the requirements of *BS EN 60439-1: 1999 Low-voltage switchgear and controlgear assemblies. Part 1. Type-tested and partially type-tested assemblies* (Regulation 712.511.1 refers).

Any junction boxes forming part of the DC side of the PV installation should be labelled to warn that parts within will remain live even after isolation from the inverter (Regulation 712.537.2.2.5.1 refers).

13.4.4 The AC side

Regulation 712.411.3.2.1.1 states that on the AC side, the PV supply cable should be connected to the supply side of any devices providing automatic disconnection for outgoing circuits supplying current-using equipment. This is normally achieved by either connecting the PV supply cable to a dedicated 'way' in the consumer unit, or to provide an additional small consumer unit for the connection of the PV power supply system specifically.

The PV supply cable must be protected against fault current by an overcurrent device at its point of connection to the AC installation (Regulation 712.434.1 refers).

If a transformerless inverter has been used, the protective conductor current of its AC circuit may at times exceed 10 mA in normal operation. Where this is the case, the relevant requirements of Regulation Group 543.7 will need to be met. See also Section 3.8 of this Guide.

If the AC PV supply cable is installed in the fabric of the building at a depth of less than 50 mm from the surface, the requirements of Regulations 522.6.201 or 522.6.202 should be considered, as appropriate. This may require the installation of an RCD meeting the requirements of Regulation 415.1.1. Further information on the additional protection of concealed cables is given in Section 11.5.7 of this Guide. Where an RCD is installed for this purpose it should disconnect all live conductors. For a single-phase installation, a double-pole RCD would be required to disconnect the line and neutral conductors.

If the RCD does not disconnect the neutral, the effectiveness of the protection no longer depends solely on the operation of the RCD, but also on the shut-down characteristics of the inverter on loss of the AC mains supply. Under G83 / 2, this could be of up to 1 second in duration, in which case the requirements of Regulations 522.6.201 or 522.6.202 may not be met.

Where the inverter of the PV installation does not provide simple separation between the AC and DC side, any RCD installed to provide fault (or additional) protection must be of a type that will operate as intended in the presence of DC currents in the residual current. Where this is the case, it will not be possible to use a type AC or type A RCD. Regulation 712.411.3.2.1.2 states that in such cases any RCD installed must be of type B to *BS EN 62423*. This requirement would not apply if the manufacturer's instructions for a particular inverter specifically state that the inverter provides galvanic isolation between the AC and DC. sides sufficient to prevent DC appearing in the AC installation.

13.4.5 Further information

Any part of the PV power supply system installation requiring maintenance, servicing or the like should be installed such that it remains accessible to allow such activities to be performed safely. (Regulation 712.513.1). This requirement would apply to any junction boxes or inverters placed in, for example, the loft of a house.

Further guidance on the installation of solar photovoltaic microgeneration systems can be found in the *NICEIC Solar PV Learning Guide.*

Anyone involved in the installation of small-scale embedded generator systems should be competent in such work in addition to their electrical installation work competency.

In addition to the standards mentioned above, reference should be made to the following sources of information as relevant to the type of system being installed:

Electrical Safety First

Best Practice Guide 3 (Issue 3: Guidance for installers and specifiers) Connecting a microgeneration system to a domestic or similar electrical installation (in parallel with the mains supply).

Energy Saving Trust

CE72 – Energy efficiency best practice in housing. Installing small wind powered electricity generating systems: Guidance for installers and specifiers.

Microgeneration Certification Scheme Microgeneration Installation Standards

MIS 3002 – Requirements for contractors undertaking the supply, design, installation, set to work commissioningand handover of solar photovoltaic (PV) microgeneration systems.

MIS 3003 – Requirements for contractors undertaking the supply, design, installation, set to work commissioning and handover of micro and small wind turbine systems.

MIS 3007 – Requirements for contractors undertaking the supply, design, installation, set to work commissioning and handover of a heating system containing a micro-cogeneration package.

CHAPTER 14 INSPECTION AND TESTING

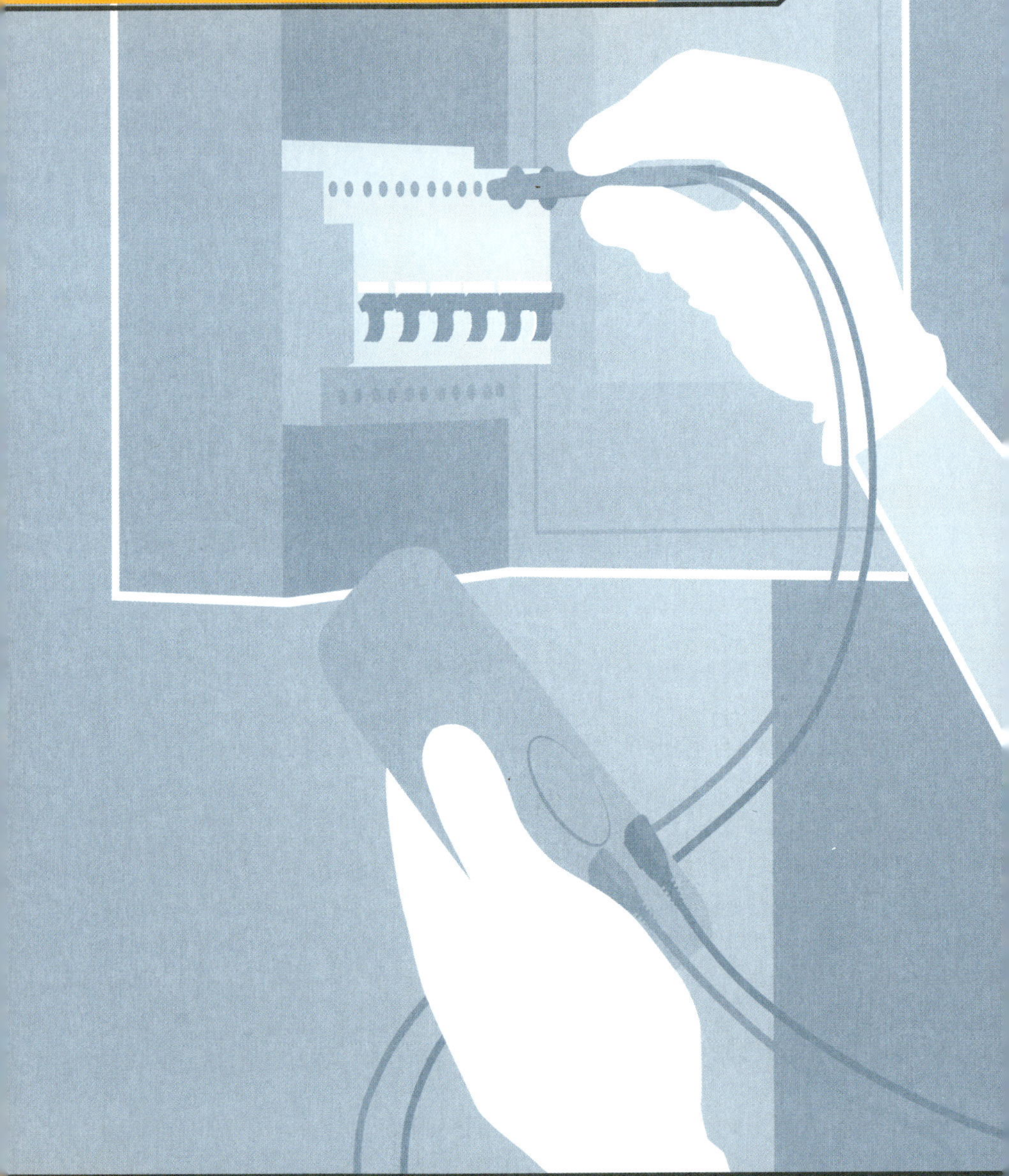

14.1 Initial verification

BS 7671 requires inspection and testing of an installation during its construction and on its completion, before a new installation is put into service or an existing installation is returned to service following additions or alterations (Regulation 641.1 refers).

Inspection forms a part of the initial verification process and should preferably be carried out before any testing (Regulation 642.1 refers), as defects or signs of significant damage might be found by the inspection that would have made the testing of the installation dangerous to the occupants of the building, or to the person(s) carrying out the inspection and testing. Furthermore, such defects or deterioration might also render any test results obtained invalid.

Wherever possible, the inspection should be carried out with the installation, or that part of it which is under scrutiny, disconnected from the supply. This is in line with the requirements of Regulation 14 of the *Electricity at Work Regulations 1989* (EWR) relating to work on or near live conductors.

The human senses of not only sight, but also touch, smell and hearing may be employed as a part of the inspection process.

Whilst the use of sight and touch require little explanation, it should be pointed out that smell and hearing can be usefully employed to identify, for example, the location of arcing at a loose connection or overheating of equipment.

Regulation 642.3 lists items that should be checked during the initial inspection of an installation, and forms the basis of both the Examples of items requiring inspection during initial verification and the Schedule of inspections for domestic and similar premises which are included in Appendix 6 of *BS 7671*.

14.2 Periodic inspection

14.2.1 General

In the case of an existing installation, Regulation 651.1 calls for periodic inspection and testing to be carried out to determine whether the installation remains satisfactory for continued use. This regulation also requires the contents of documentation such as the original Electrical Installation Certificate, or previously issued Periodic Inspection Reports relating to the installation, where available, to be taken into consideration.

Regulation 651.1 further states that where no such previous documentation is available an investigation of the electrical installation should be carried out prior to performing any inspection and testing activities.

The periodic inspection should be performed without dismantling, or the minimum of dismantling (Regulation 651.2 refers), in order to reduce the risk of the inspection process actually introducing a fault.

When carrying out a periodic inspection of an existing installation, it is not always possible or necessary to carry out the same degree of inspection and testing as for new installation work. Furthermore, it may not be practicable or indeed physically possible to inspect and test the whole of the installation.

This is recognised by *BS 7671*, which contains a number of requirements relating to the extent of the periodic inspection and testing of an installation, and limitations relating to various aspects of the inspection and testing process.

14.2.2 Details of the installation to be inspected

Regulation 653.2 includes a list of items that shall be recorded on an Electrical Installation Condition Report (EICR). Where for operational reasons it is not possible to test the complete installation, the item marked; Details of the electrical installation covered by this Report, and shown in Fig 1, details those parts of the installation that were inspected and tested

The details of the existing installation to be covered by a periodic inspection should be agreed in advance between the person carrying out the inspection and testing and the client.

A number of factors must be taken into consideration when deciding upon the details to be inspected and tested, including:

- the size of the installation,
- the wishes of the client,
- the requirements of third parties, such as insurance providers and licensing authorities,
- the usage of the building and ambient environmental conditions,
- the age and general condition of the electrical installation,
- whether or not it is acceptable for all, or parts of, the installation to be isolated from the supply,
- acceptable durations of any agreed isolation of supply,
- mutually agreeable times for the inspection and testing to be carried out,
- the time elapsed since the initial verification was completed, or the last periodic inspection was carried out,

- the availability of records relating to the installation, maintenance and previous inspections of the installation
- the effectiveness of any on-going maintenance as evidenced during the inspection process.

This information has to be described in the EICR and is issued on completion of the inspection and testing process, as required by Regulation 653.1.

14.2.3 Limitations on the inspection and testing

Regulation 653.2 also requires any limitations of the inspection and testing to be detailed on the EICR (see also Fig 1).

Any limitations should be agreed with the client prior to work starting. However, it is important that neither the client nor the person carrying out the work imposes limitations unnecessarily, as this would result in the Report having less validity.

The following are typical examples of limitations.

i) No testing at height where this presents accessibility problems (where, for example, temporary access equipment would be required).

ii) Particular circuits or parts of an installation not tested where disconnection of supply would cause an unacceptable level of disruption to the operations of the client.

iii) Particular circuits or parts of an installation not tested where the implications of an unwanted disconnection would be prohibitively expensive in terms of lost data etc.

Items ii) and iii) above are not generally applicable to installations in domestic premises.
It should be noted that the following testing activities do not have to be listed as limitations.

- Carrying out an insulation resistance test on all circuits connected to a consumer unit or distribution board collectively (Regulation 643.3.1 refers).
- Carrying out an insulation resistance test where all line conductors are connected together and tested to the earthing arrangement where a circuit includes electronic devices that might otherwise be damaged or have an influence on the test results obtained (see Regulations 643.3.1 and 643.3.2).
- Carrying out an insulation resistance test at 250 V DC on circuits where surge protective devices or other equipment are likely to be present, which might otherwise be damaged as a result of the testing or have an influence on the test results obtained, subject to a minimum acceptable insulation resistance of 1 MΩ being achieved (Regulation 643.3.2 refers). Such items of equipment should be disconnected prior to carrying out the test where this is practicable.

Although the above items do not need to be listed as limitations of the inspection and testing process, their use is nevertheless noteworthy and so an appropriate description of such should be included in the 'Remarks' column against the appropriate circuit on the Schedule of Test Results.

The details of the installation covered by the Report might be the whole of the installation in the premises, or it might be only a part of that installation. The part could, for example, be a particular area (such as the second floor only) or particular circuits (such as all lighting circuits). The details could also be indicated by exclusions from the whole electrical installation, such as the central heating system wiring or the fire alarm system. The wording will need to be tailored exactly to each inspection requirement and, in some instances, a continuation sheet (appropriately numbered, and identified) may be necessary.

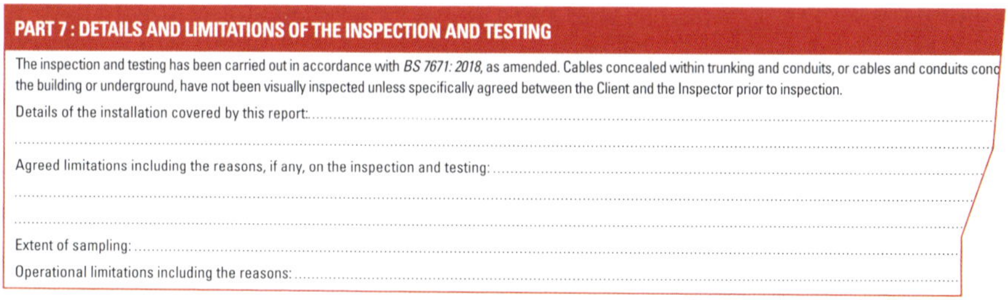
PART 7 : DETAILS AND LIMITATIONS OF THE INSPECTION AND TESTING

The inspection and testing has been carried out in accordance with *BS 7671: 2018*, as amended. Cables concealed within trunking and conduits, or cables and conduits conc the building or underground, have not been visually inspected unless specifically agreed between the Client and the Inspector prior to inspection.

Details of the installation covered by this report:...............................

Agreed limitations including the reasons, if any, on the inspection and testing:...............................

Extent of sampling:...............................

Operational limitations including the reasons:...............................

Fig 1 Part 7 of the EICR showing the details, limitations, and extent of sampling of the installation

A periodic inspection must include a thorough visual inspection of all the electrical equipment that comprises the extent of the installation described previously, and which is not concealed or inaccessible (including electrical equipment installed in lofts that requires maintaining).

14.2.4 Sampling

In most cases, it will be appropriate to apply a sampling process to the detailed inspection of the internal condition of equipment, the condition of joints and terminations etc.

If sampling is intended, this should be made clear to the client, and the initial degree of sampling for both the inspection and the testing elements should be agreed before the work commences. Should the actual degree of sampling carried out be different from that agreed initially (because, for example, increased sampling was necessary due to deficiencies being found during the initial sampling), the actual degree should be recorded on the Report.

Details of the sampling technique and identification of the equipment selected for inspection and testing on a sampling basis should also be fully identified, on an additional page if necessary.

CHAPTER 15
TESTING

15.1 Introduction

Electrical testing inherently involves a degree of risk. Persons carrying out electrical testing have a duty to ensure the safety of themselves and others from any possible danger resulting from the tests or from access to live parts. This requires strict adherence to safe working practices, such as the 'Isolation procedure' (see Appendix B), and performing appropriate checks to confirm that:

- the test equipment and leads are safe and suitable for the intended purpose,
- the equipment to be tested is safe for the tests, and
- the working environment does not impair safety.

Measuring instruments, monitoring equipment and methods used for electrical testing should conform to the relevant parts of *BS EN 61557-16: 2015 Electrical safety in low voltage distribution systems up to 1 000 V AC and 1 500 V DC. — Equipment for testing, measuring or monitoring of protective measures.*

Other measuring instruments are not precluded providing they offer similar levels of safety and performance (Regulation 643.1 refers).

15.2 Safety

Persons carrying out testing, should be aware of the danger present when connecting test instruments to live parts and take suitable precautions as required by the *Electricity at Work Regulations 1989* (EAWR). General advice on such safety matters is given in HSE Guidance Note GS38: *Electrical test equipment for use on low voltage electrical systems.*

GS38 is available as a free download from: **http://www.hse.gov.uk/pUbns/priced/gs38.pdf**

Particular care should be exercised where the installation is connected to additional supply sources, such as those provided by standby generators and uninterruptible power supplies

This chapter gives guidance on the testing of domestic electrical installations. Where more detail is required on inspection and testing techniques and on how to complete the necessary certificates and schedules of inspection and schedules of test results, the following publications are available:

1. Inspection, Testing and Certification – including Periodic Reporting
2. A to Domestic Periodic Inspection, Testing and Reporting

These publications can be purchased on-line from **www.shop.niceic.com**

Regulation 641.1 calls for testing (and inspection) to be carried out, as appropriate, at a number of occasions during the construction of the installation.

In many cases there may be a significant time interval between the initial placement of conductors (sometimes referred to as 'first fixing') and the final handover of the installation. It is possible for damage to have occurred during this time when, for example, cables were plastered over and flooring and ceilings were installed.

It is essential, therefore, that all the relevant tests are performed as part of the final verification of the electrical installation. In some cases, this will mean repeating a test carried out previously. It is not acceptable to rely solely on test results that were obtained at an earlier stage of the installation process.

15.3 Sequence of tests – initial verification

The sequence of testing shown in Table 15.1 is appropriate when carrying out the initial verification of circuits within a typical domestic installation.

Table 15.1 Initial sequence of tests

a)	Before the supply is connected	Regulation
i)	Continuity of protective conductors	643.2
ii)	Continuity of ring final circuit conductors	
iii)	Insulation resistance	643.3
iv)	Polarity by continuity methods	643.6
v)	Earth electrode resistance (where fitted)	643.7.2
b)	**With the electrical supply connected**	
vi)	Confirm polarity before further testing (live test)	643.6
vii)	External earth fault loop impedance	643.7.3
viii)	Prospective fault current at the origin (both short-circuit and earth fault)	643.7.3.201
ix)	Tests to verify operation of RCD/RCBO providing fault protection Tests to verify operation of RCD providing additional protection	643.7.1 643.8
x)	Tests to verify operation of AFDDs where fitted	643.10
xi)	Any functional testing deemed appropriate	643.10

15.4 Sequence of tests – periodic inspection

Whilst no sequence of tests is stated for periodic testing, the sequence shown in Table 15.2 would be appropriate when carrying out a periodic inspection of circuits within a typical domestic installation.

Table 15.2 Suggested sequence of test for a periodic inspection

	Test	Regulation
i)	External earth fault loop impedance or Earth electrode resistance	643.7.3 643.7.2
ii)	Prospective fault current at the origin (both short-circuit and earth fault)	643.7.3.201
iii)	Continuity of protective conductors and continuity of ring final circuit conductors	643.2
iv)	Insulation resistance	643.3
v)	Polarity by continuity methods	643.6
vi)	Earth fault loop impedance of circuits	643.7.3
vii)	Tests to verify operation of RCD/RCBO providing fault protection Tests to verify operation of RCD providing additional protection	643.7.1 643.8
viii)	Tests to verify operation of AFDDs where fitted	643.10
ix)	Any functional testing deemed appropriate.	643.10

Test Procedures

Test instrument

The test instrument to be used for continuity is an ohmmeter having a low resistance range, or insulation and continuity test instrument set to the 'continuity' range. Typically the use of a multi-meter is not permitted as they are incapable of providing sufficient current or voltage required for the tests.

Continuity test readings of less than 1 Ω are common, therefore, the resistance of the test leads can be significant, and should not be included in any recorded test result. Prior to commencing the continuity test, the test leads should be nulled (see Fig 15.1). This is particularly important when using the 'wander' lead method for determining R_2. This method will be mentioned later in the chapter.

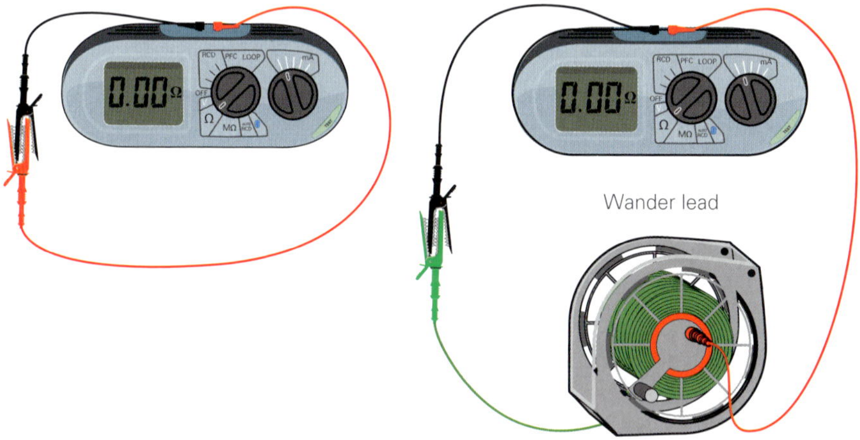

Fig 15.1 Nulling the resistance of the test leads

The nulling process varies between different makes of instruments; therefore, it is important that manufacturers' instructions are consulted. Most modern test instruments have an 'auto-null' facility so that the test value recorded does not include the resistance of the test leads. Failing that, make a note of the test leads resistance so that figure can be subtracted from the measured test values.

15.5.1 Continuity of protective conductors

Measurement errors resulting from parallel paths

Before carrying out a test to confirm the continuity, or to measure the resistance, of a protective conductor, care should be taken to locate and disconnect (where practicable) any conductive paths which may be connected in parallel with all or part of it (see Fig 15.2). Such parallel paths could be present due, for example, to a connection from a protective conductor to an electrically continuous system of pipework at one or more points along the protective conductor. Such parallel paths could lead to incorrect test results and may even conceal the fact that a protective conductor is not electrically continuous.

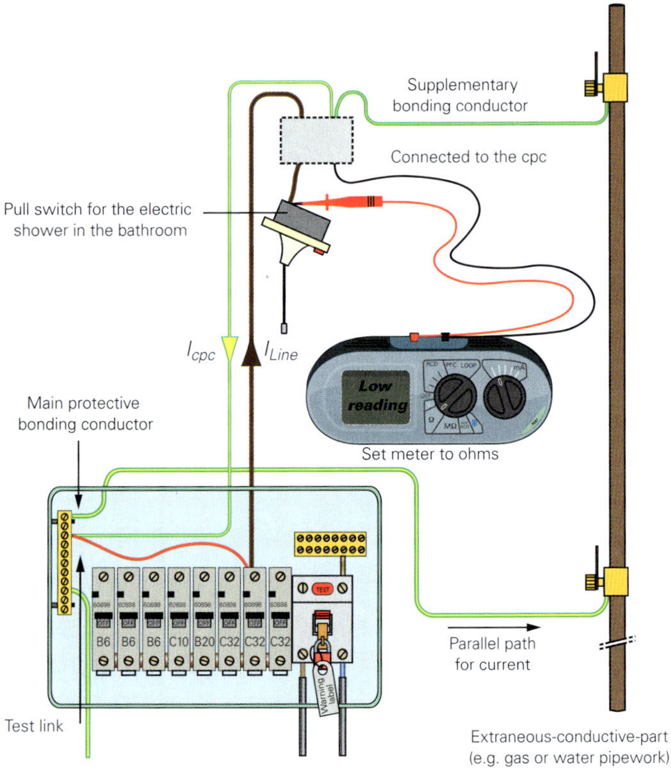

Fig 15.2 Example showing how a parallel path might exist

Reason for testing the continuity of protective conductors

It is essential to confirm that all circuit protective conductors and bonding conductors are continuous. Otherwise, an exposed-conductive-part, an extraneous-conductive-part or the earthing terminal of a point or accessory could be left without an effective connection to Earth, giving no fault protection in the event of an earth fault.

(R_1 + R_2) method

This method is applicable to circuit protective conductors and their associated line conductors. It is not applicable to main or supplementary bonding conductors, which require the 'wander' lead or R_2 method to be used. Fig 15.3 shows how the (R_1 + R_2) method is performed on a lighting circuit.

The $(R_1 + R_2)$ method of testing the continuity of circuit protective conductors has the following advantages:

1. The information required to complete the $(R_1 + R_2)$ column of the Schedule of Test Results is obtained directly.
2. Polarity is verified as each continuity test is performed.
3. The method may be more convenient than the R_2 'wander lead' method (see following section).

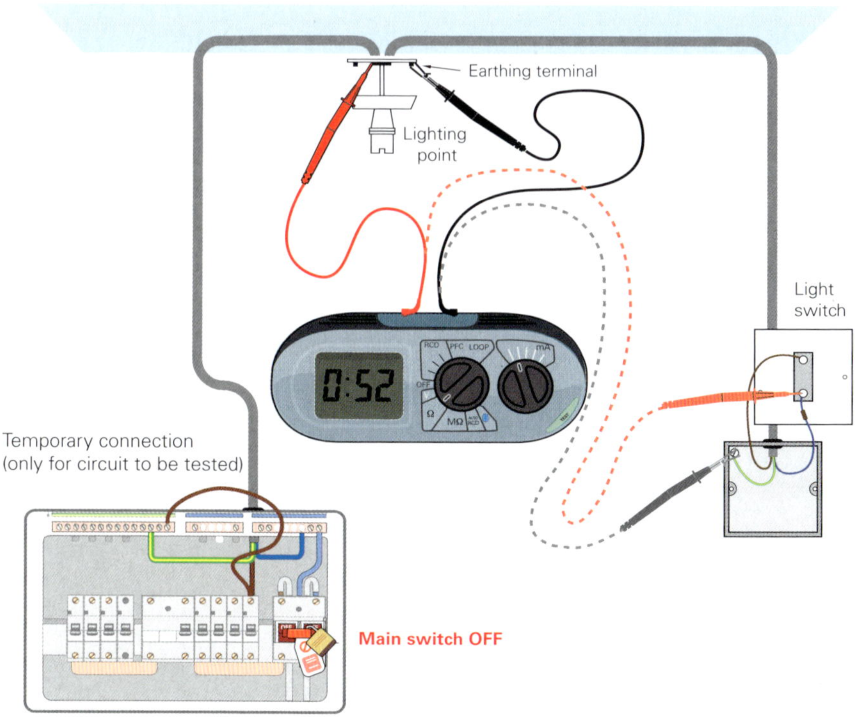

Fig 15.3 Continuity testing using the $(R_1 + R_2)$ method

The test provides a measured value of resistance for the line conductor and the circuit protective conductor of a circuit, in series. The test may be carried out by temporarily connecting together the line conductor and circuit protective conductor at the supply end, and then connecting the test instrument to the terminals of the switch line and the circuit protective conductor at each point and accessory on the circuit in turn. By operating the switch, the polarity can be verified at the same time.

The value of (R_1 + R_2) measured at the point or accessory electrically most remote electrically from the supply will be the maximum value, and it is this value that should be recorded on the Schedule of Test Results. The test readings at the other points and accessories on the circuit serve to verify that the protective conductor is continuous to each of them, and that the polarity is correct.

R_2 method ('wander' lead)

This method is used principally for testing main and supplementary protective bonding conductors. Using a long lead (wander lead), the leads of the instrument are attached as shown in Fig 15.4, having temporarily disconnected the conductor from the MET to remove parallel paths. To confirm a sound low resistance connection between the bonding clamp and the pipe, the test connection should be made to the pipe and not the conductor.

This method may also be used to verify the continuity of earth connections made at every position in a circuit, such as at the earth points of socket-outlets, switches, lighting points, fixed equipment and the like. The wander lead is connected to the main earthing terminal and a lead of the test instrument is connected to the other end of the wander lead. The remaining lead of the test instrument makes contact to each earth point within the circuit. Provided that no parallel paths are present, an accurate measurement of continuity of the circuit protective conductor back to the main earthing terminal is obtained.

The wander lead used to carry out this test may be as long as 50 m in length; therefore, the resistance of the lead will be significant. It is essential therefore that the resistance of the wander lead is either 'nulled' prior to testing (see Fig 15.1) or is subtracted from the value obtained from testing.

15.5.2 Continuity of ring final circuit conductors

Incorrectly connected or open circuit, conductors in an installed ring final circuit can lead to overloading of the circuit cables. Testing by the above procedure verifies continuity and correct connection, as well as correct polarity.

Testing also provides confirmation that both legs of the ring final circuit have been connected into the same protective device. The incorrect connection of a ring final circuit across two protective devices would leave the installation in a dangerous condition particularly when safe isolation of a final circuit was required.

The cables of a ring final circuit start at the outgoing terminals of a protective device within a consumer unit or distribution board, connect to all the points in the ring, and return to the same outgoing terminals. The live (line and neutral) conductors must form a complete unbroken loop without interconnections, as must the circuit protective conductors, unless it is formed by a continuous metal covering or enclosure (such as a steel conduit) containing all the live conductors.

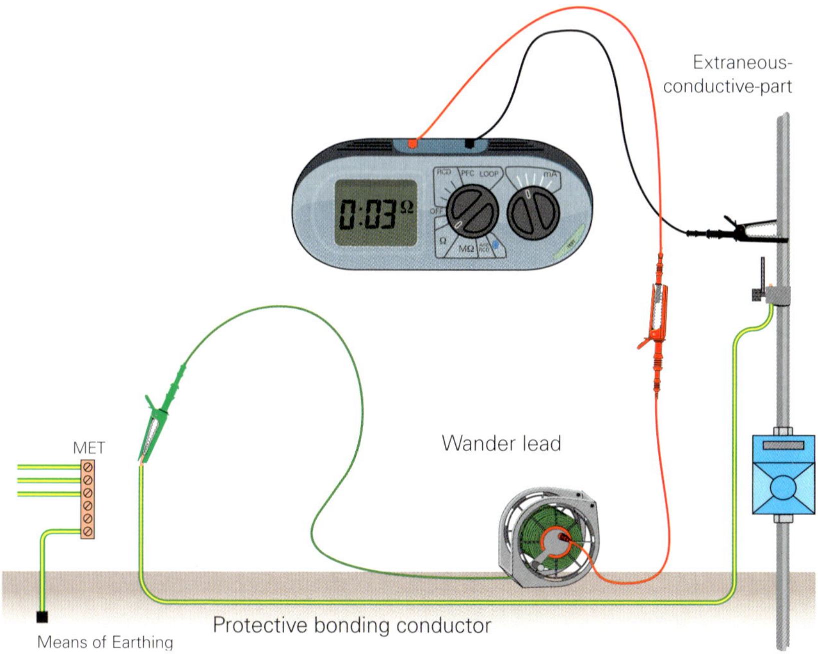

Fig 15.4 The 'wander lead' method to measure the continuity of a main protective bonding conductor

Procedure:

Step 1

Confirm that the circuit to be tested has been securely isolated.

Null the test leads or make a note of their resistance.

Using a low resistance ohmmeter; set it on its lowest scale (typically 0-2 Ω or similar).

Measure the end-to-end resistance of the:

1. line conductor (r_1)
2. neutral conductor (r_n)
3. cpc (r_2)

(see Fig 15.5)

The results obtained should be more or less equal (± 0.05 Ω, say) except where the circuit is wired using 'twin and earth' type cables having a cpc of reduced cross-sectional area.

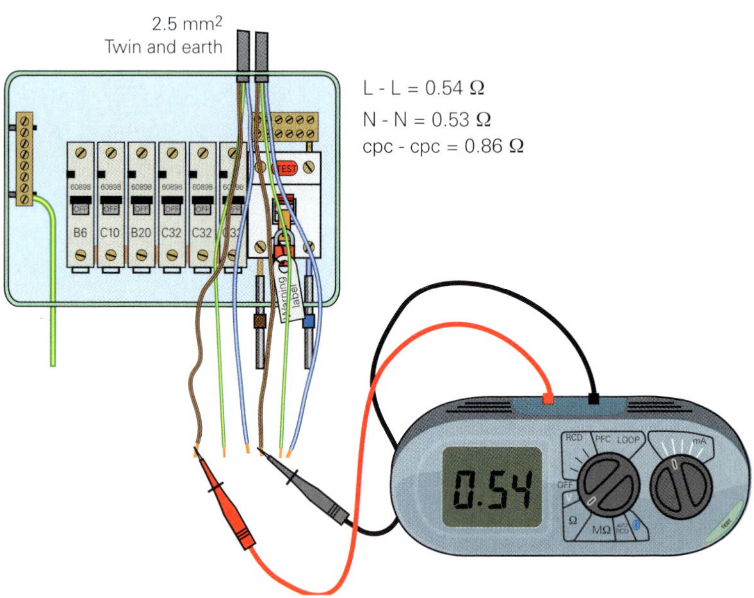

Fig 15.5 Step 1 of ring final circuit continuity test

Step 2

Interconnect the line and neutral conductors at the consumer unit as shown in Fig 15.6. Measure the resistance between line and neutral terminals at every socket-outlet connected to the circuit. The results obtained should be substantially the same at each socket-outlet and should be approximately one quarter of the combined resistance of the line loop (r_1) and neutral loop (r_n) that is; meter reading $= \frac{r_1 + r_n}{4}$

For the example values given in Fig 15.5 the reading at each socket-outlet:

$$\frac{r_1 + r_n}{4} = \frac{0.54 + 0.53}{4} = 0.267\Omega$$

Any socket-outlets connected as unfused spurs will have a higher resistance value as a result of the resistance of the spur conductors.

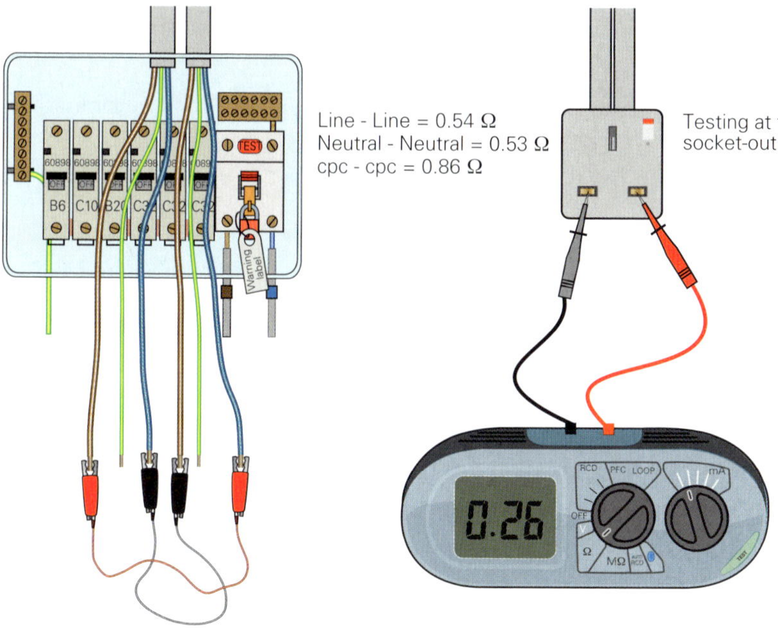

Fig 15.6 Step 2 of ring final circuit continuity test

Step 3

Interconnect the line and cpc conductors of the circuit at the consumer unit. Measure the resistance between line and cpc terminals at every socket-outlet connected to the circuit under test. The results obtained should be substantially the same at each socket-outlet and should be approximately one quarter of the combined resistance of the line loop (r_1) and cpc loop (r_2). For the example values given in Fig 15.5 the reading at each socket-outlet would be;

Socket-outlets wired as spurs will give a slightly higher resistance reading depending on the length of the spur. $\frac{r_1 + r_2}{4} = \frac{0.54 + 0.86}{4} = 0.35\ \Omega$

The highest value of resistance obtained should be recorded on the Schedule of Test Results as the ($R_1 + R_2$) value of the circuit.

15.5.3 Insulation resistance

General

Carrying out an insulation resistance test gives an indication of the condition of the insulation of conductors and any equipment to which they are connected. Effective insulation is necessary to provide basic protection and to prevent short-circuits and earth faults.

The insulation resistance of a number of parallel circuits connected to a consumer unit may be tested collectively. It is acceptable to carry out an insulation resistance test of a distribution circuit (if any), the consumer unit and the final circuits originating from it, with the protective devices of the final circuits in the ON position, and the current-using equipment temporarily disconnected (Regulation 643.3.2 refers). The insulation resistance shall not be less than the appropriate value given in Table 64 of *BS7671*.

Testing

Preconditions and precautions

Before testing can commence it is essential that:

- All pilot and indicator lamps are disconnected from circuits. Typically, neon indicators have a resistance of about 200 kΩ and could lead to inaccurate and misleading readings
- Disconnect capacitors and/or inductors as they may also give inaccurate readings
- Disconnect or switch off all connected loads, or the winding of the load(s) and not the insulation resistance will be tested
- All voltage sensitive devices, such as dimmer switches and electronic timers must be disconnected.
- The earthing conductor must be connected.

Where the above bullet pointed items is impracticable or where doubt exists as to whether all such equipment has been disconnected, the following options are available:

1. Insulation resistance may be measured between live conductors and protective conductor(s) with all line and neutral conductors connected together. Where this method is used, there is no potential difference between the line and neutral terminals of connected equipment or the components therein.

2. The test voltage applied may be reduced to 250 V DC, but the minimum acceptable value of insulation resistance remains 1 MΩ (Regulation 643.3.2 refers).

Testing of low voltage circuits and FELV circuits

Temporarily disconnect all current-using equipment. Remove any lamps from lampholders. Place all switches in the ON position. Two-way and, where fitted, intermediate, switches will need to be operated during the test to ensure that all live conductors are tested.

Confirm that the circuit to be tested has been de-energised. Then, using a high resistance ohmmeter set to 500 V DC and for the highest resistance scale on the instrument; carry out the following test procedure and record the results obtained.

For a single-phase installation, test between line and neutral as shown in Fig 15.6

Then test between live conductors and protective earth connected to the means of earthing, as shown in Fig 15.7.

However, where the consumer unit has been connected to the energy meter, the test must be carried out on the load side of the main switch in the OFF/isolation position, but with all the circuit protective devices in the ON position.

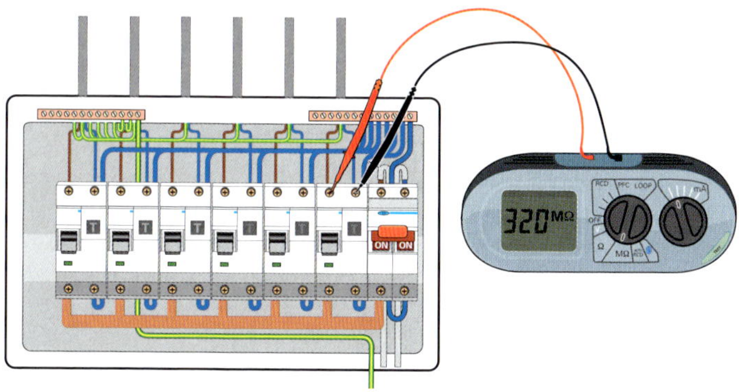

Fig 15.7 Insulation resistance test between line and neutral

Where a consumer unit contains devices such as RCDs/RCBOs, extra care must be exercised to ensure collective testing is appropriate as these devices can give a poor insulation resistance reading and they are susceptible to damage from the 500 V DC test voltage. Where necessary, circuits protected by such devices should be tested individually on the load side.

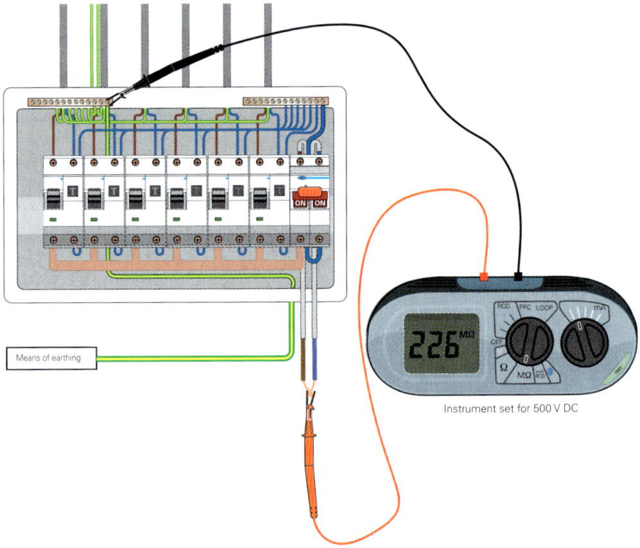

Fig 15.8 Insulation resistance between live conductors and earth

When performing an insulation resistance test between live and protective conductors, protective conductors must be connected to the earthing arrangement, that is, the general mass of Earth to comply with Regulation 643.3.1.

The insulation resistance of a FELV circuit, when tested, should meet all of the requirements applicable to a low voltage circuit (Regulation 643.3.2 refers).

The minimum acceptable value of insulation resistance for a distribution circuit with a distribution board or consumer unit and all its final circuits connected is 1 MΩ (See Table 64 of *BS 7671*).

Testing of SELV, PELV and FELV circuits

Where the protective measure for a circuit is SELV or PELV, it should be confirmed by an insulation resistance test using an applied test voltage of 500 V DC, that the live parts so protected are sufficiently separated from the live parts of any other non SELV or PELV circuits within the installation (Regulation 643.4 refers).

The minimum acceptable value of insulation resistance when confirming the separation of the conductors of SELV and PELV circuits from those of non-SELV or PELV circuits in the same location is 1 MΩ (Regulation 643.3.2 refers).

The minimum acceptable value of insulation resistance when confirming the separation of the conductors of SELV and PELV circuits from each other, and from Earth, is 0.5 MΩ at a test voltage of 250 V DC. (Table 64 of *BS 7671* refers).

The insulation resistance of a FELV circuit, when tested shall be at the same test voltage as that applied to the primary side of the source and shall meet all the test requirements for low voltage circuits (Regulation 643.3.2 refers).

15.5.4 Polarity

By continuity methods (installation or circuit de-energised)

It should be noted that a great deal of the verification of polarity may be done by inspection. Furthermore, much of the verification of polarity can be carried out at the same time that the continuity testing is being done. In the case of ring final circuits, the continuity of the cpc is verified as a part of the ring final circuit continuity test.

Using a low resistance ohmmeter set on its lowest scale (typically 0-2 Ω or similar) carry out the continuity testing in accordance with the $(R_1 + R_2)$ method described in Section 15.5.1 of this chapter, paying particular attention to confirming that any installed single-pole switching devices are correctly connected in the line conductor only.

Other items to be confirmed are that:

1. fuses and single-pole circuit-breakers are connected in the line conductor only,
2. lampholders (except for those designed for E14 and E27 lamp caps) having a centre contact have the outer or screwed contact connected to the neutral conductor, and
3. all accessories, such as socket-outlets and fused connection units, have been correctly connected (Regulation 643.6 refers).

Verifying the polarity of an incoming supply (or an existing live circuit)

To confirm that, in service, the polarity of an installation is correct in addition to employing the techniques previously described to check the wiring of the circuit(s) the polarity of the incoming supply should be verified.

It should be noted that when checking the polarity of the incoming supply different procedures are required for TT or TN-S and TN-C-S systems.

For a TT or TN-S system, a two-pole voltage indicator or test lamp can be used and it should indicate a live supply when connected across L-N and L-E but not when connected across N-E.

For further details see Chapter 8 of the NICEIC and ELECSA publication, *Inspection, Testing and Certification*.

15.5.5 Protection by automatic disconnection of the Supply (ADS)

ADS, is the most common method used to provide fault protection for circuits forming part of TN or TT systems, and may be achieved by either of the following devices:

1. An overcurrent protective device
2. An RCD

To verify that devices installed for fault protection will provide ADS in accordance with *BS 7671*, the earth fault loop impedance path, Z_S, should be determined for every distribution circuit and final circuit that relies on ADS as its protective measure against electric shock (Regulation 643.7 refers).

Where ADS is provided by an overcurrent protective device, Z_S measurements are made to verify that the maximum earth fault loop impedance permitted for the device to achieve the required disconnection time is not exceeded. (Measurement of earth fault loop impedance, (Z_S) is covered later in this chapter.)

Where an RCD is used to provide fault protection the maximum values of Z_S given in Table 41.5 are applicable. Guidance on the verification of RCDs used to provide fault protection is covered later in this chapter under 'Operation of RCDs'.

It should be noted that irrespective of whether an installation forms part of a TN or TT system, where an RCD is used to provide fault protection the circuit should also incorporate an overcurrent protective device in accordance with Chapter 43 of *BS 7671* (Regulations 411.4.5 and 411.4.201 refer).

As part of the verification for ADS, the means of earthing for the installation should be confirmed before the installation is energised. For this purpose the external earth fault loop impedance Z_e for an installation is required to be determined, by enquiry or by measurement, and recorded on the certificate or report.

15.5.6 Measurement of external earth fault loop impedance (Z_e)

It is necessary to record the external earth fault loop impedance, Z_e, in the Supply characteristics and earthing arrangements section of the Electrical Installation Condition Report.

Although Z_e may be determined by enquiry or calculation, a measurement of Z_e should still be undertaken as part of the initial verification to verify that the intended means of earthing is present and its impedance is suitable for the type of earthing arrangement.

Before the test, the earthing conductor must be disconnected from the main earthing terminal or otherwise separated from all parallel earth paths such as bonding and circuit protective conductors. Any parallel earth paths present during the test would result in a false reading of Z_e being obtained and might perhaps conceal the fact that the intended means of earthing is defective, or even non-existent.

For safety reasons, it is essential that the entire installation is isolated from the supply before the earthing conductor (or, if necessary, any other protective conductors) are disconnected from the main earthing terminal, and that the installation remains isolated for the duration of the test.

On completion of testing and before the installation is re-energised, the earthing conductor or other protective conductors that were disconnected to facilitate the measurement of Z_e must be reconnected.

The inspector should take particular care during the testing process, as typically fault conditions are most severe at the origin of the installation, where this test is being performed.

The earth loop impedance test instrument (or test range on a multi-test instrument) should be used to carry out the earth loop impedance test.

In the two-lead version, connection is made between line and earth of the incoming supply. In the three-lead version it is necessary to make a connection to the neutral of the incoming supply as well as to line and earth.

Where the supply is provided with an earth connection from the electricity distributor's network, such as from a TN-S or TN-C-S system, it is unlikely that Z_e will exceed 1 ohm, and the lowest range of the instrument will be the most appropriate (usually 0 to 20 Ω).

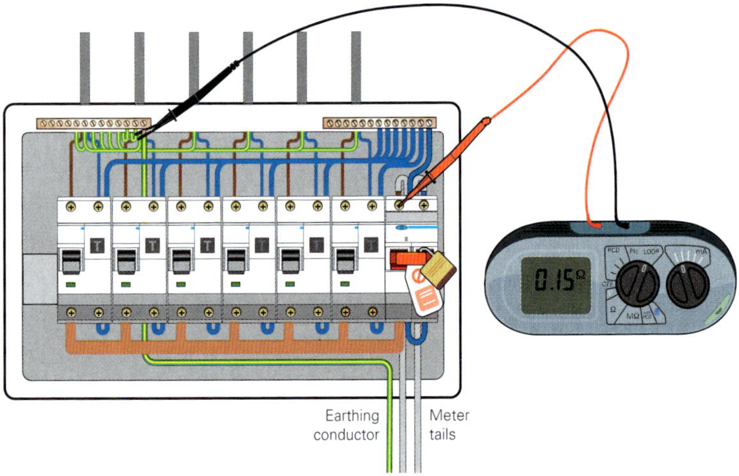

Fig 15.9 External earth fault loop impedance

Typical maximum values of Z_e for single-phase supplies of up to 100 A published by electricity distributors are:

- TN-S 0.80 Ω
- TN-C-S 0.35 Ω

Note: The typical values given for Z_e may vary depending on the location and should be confirmed with the appropriate DNO.

Earth electrode resistance

General

An installation forming part of a TT system is required to be equipped with an installation earth electrode as its means of earthing. The function of the earth electrode is to make electrical contact with the general mass of Earth to provide a suitable path for earth fault current to return to the source (such as via the earth electrode at the distribution transformer).

Measurement of the resistance of the installation earth electrode to the general mass of Earth provides information necessary for verifying that the earthing arrangement satisfies the requirements of both *BS 7671* and the particular installation design. The measured resistance value needs to be recorded, together with the type of electrode and its location, in the appropriate boxes in the 'Particulars of the installation' section of the Electrical Installation Certificate and of the Domestic Electrical Installation Condition Report.

Details on how to perform testing of earth electrodes are given in NICEIC publication *Inspection, Testing and Certification*.

15.5.7 Prospective fault current at the origin

BS 7671 requires that the prospective fault current is determined at every relevant point within the installation (Regulation 643.7.3.201 refers). Regulation 434.5.1 indicates that unless another device installed upstream provides the necessary back up protection, the rated short-circuit breaking capacity of a fault current protective device must not be less than the maximum prospective fault current at the point at which it is installed.

The maximum prospective fault current forms another of the supply characteristics and earthing arrangements that need to be recorded in the 'Nature of supply parameters' section of the Electrical Installation Certificate and the Electrical Installation Condition Report.

It should be noted that testing of maximum prospective fault current is not required where a consumer unit to *BS EN 61439-3* is installed in a domestic (household) or similar premises, and the maximum prospective fault current declared by the distributor is 16 kA. Given such circumstances, 16 kA should be recorded on the certificate or report.

Where measurement is required in a single-phase installation, the prospective fault current is the greater of either the fault current between:

- the line conductor and neutral conductor (prospective short-circuit current), or
- the line conductor and Earth (prospective earth fault current).

The highest reading obtained from these tests should be recorded as the value of I_{pf}.

Details on how to perform prospective fault current testing are given in NICEIC publication *Inspection, Testing and Certification.*

15.5.8 Circuit earth fault loop impedance (Z_s)

Where it is safe to do so, measured values of Z_S can be obtained wholly by means of an earth fault loop impedance test instrument. However, before a measurement of Z_S is undertaken continuity of protective conductors should be verified (Regulation 643.7.2 refers).

The purpose of determining the circuit earth fault loop impedance is to verify that the protective devices for distribution and final circuits will operate within the required time

For each circuit recorded on the NICEIC or ELECSA combined schedule of 'Circuit details and test results' both the maximum tabulated value of Z_S, from the appropriate table contained in Chapter 41 of *BS 7671*, and the maximum measured (or calculated) value of Z_S should be recorded in the relevant column. Where the tabulated value is obtained from a source other than *BS 7671*, the details of the source should be identified at the 'foot' of the schedule.

This test is conducted with the supply 'ON', so all necessary precautions must therefore be taken to prevent danger when conducting live testing. Testing should only be carried out after a thorough inspection and continuity testing have confirmed, so far as is reasonably practicable, that all circuit protective conductors, and main and supplementary bonding conductors, are in place. If this procedure is not observed, there is a possibility that exposed-conductive-parts and extraneous-conductive-parts may become live during the testing process, thus exposing persons and livestock to the risk of electric shock.

For lighting circuits, connect one of the instrument lead to the earth terminal before connecting the second lead to the line or switch line terminal at the lighting enclosure or ceiling rose at the electrically furthest point or accessory (see Fig 15.11). Perform the test and then switch off the circuit on completion of the testing.

As illustrated in Fig 15.10 and Fig 15.11, measurements of Z_S should be made at the furthest point of every distribution circuit and final circuit where fault protection is provided by automatic disconnection of supply (ADS).

For testing socket-outlets, a proprietary test lead with a fitted plug should be used. As a minimum, testing should be performed on the outlet which is situated electrically furthest from the origin of the circuit. Where this is not known, test all socket-outlets connected to the circuit and record the highest value on the test schedule.

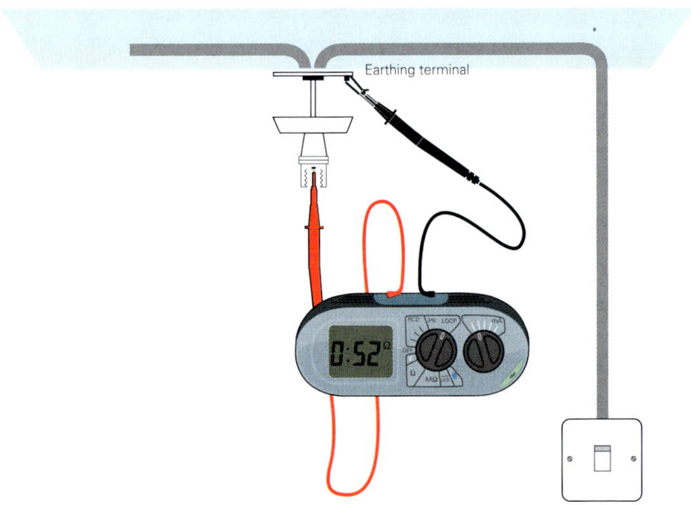

Fig 15.10 Earth loop impedance test on a lighting circuit

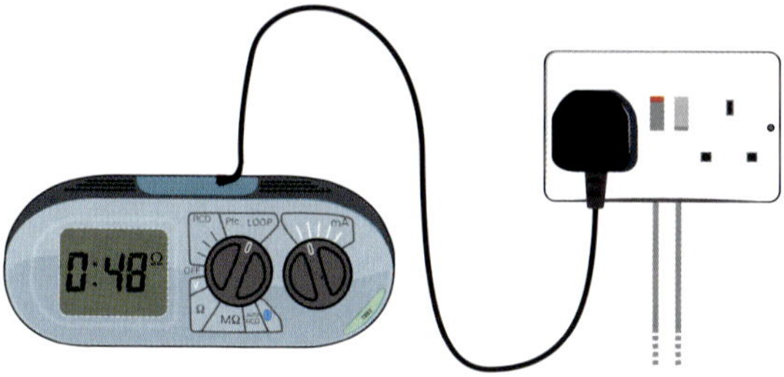

Fig 15.11 Earth loop impedance test on a socket-outlet circuit

For fault protection, the limiting values of earth fault loop impedances, Z_S, for many commonly-used overcurrent protective devices are given in Chapter 41. However, it is preferable to make use of actual manufacturer's data.

The values given by manufacturers and in Chapter 41 are the limits that apply under earth fault conditions, when the temperature of the conductors can be expected to be higher than when testing is undertaken (usually under no-load conditions). Consequently, the values of earth fault loop impedance when measured at ambient temperature should be lower than the limits set out in *BS 7671*.

It is generally accepted that, where the measured earth fault loop impedance of a circuit is not greater than 80 % of the relevant limit specified in *BS 7671*, the impedance can be expected to be sufficiently low under earth fault conditions to meet the relevant limit specified in *BS 7671*, and for the protective device to automatically disconnect within the time specified.

Table 41.5 of *BS 7671* states the maximum acceptable earth fault loop impedances to ensure RCD operation where an RCD is used for earth fault protection (see Regulation 411.4.204 for TN systems and Regulation 411.5.3 for TT systems).

15.5.9 RCDs

Note: The detail given in this section will apply equally to RCBOs conforming to *BS EN 61009-1* (Regulation 411.4.204 refers).

RCDs with a rated residual operating current not exceeding 30 mA may be used to provide additional protection in the event of failure of basic and/or fault protection or carelessness by the user (Regulation 415.1.1 refers). Where this is the case each RCD should be inspected and tested to confirm that the relevant requirements of Chapter 41 of *BS 7671* are met. Suitable methods of verifying the effectiveness of RCDs used to provide additional or fault protection are outlined later in this chapter.

Operation of RCDs

To verify the effectiveness of an RCD used to provide automatic disconnection in the event of a fault as required by Regulation 643.7.1 or to verify the effectiveness of an RCD used to provide additional protection as required by Regulation 643.8 the following inspections and tests are applicable:

- a visual inspection to confirm that the RCD is of an appropriate type, rated current or current setting, and sensitivity for its intended application,
- testing of the RCD using a suitable test instrument according to *BS EN 61557-6*, and
- operation of the integral test facility to confirm correct function.

An RCD may also be used for protection against fire. An RCD that satisfies the requirements for ADS is considered to satisfy the requirements for providing protection against fire, and should comply with Regulations 531.3.1 to 531.4.

Where an RCD is employed to achieve the disconnection time required by Table 41.1 it is first necessary to verify that the maximum earth fault loop impedance (Z_s) stated for a particular sensitivity of RCD in Table 41.5 are not exceeded for the circuit in which they provide protection.

RCD testing

The purpose of the RCD test is to confirm that the device will operate within the disconnection time required by *BS 7671*.

Where an RCD is used to provide fault protection, the test should be performed at any current equal to or greater than the rated residual operating current of the device. The disconnection time should comply with the general requirements detailed in Chapter 41.

Where an RCD is used to provide additional protection, the test should be performed at any current equal to or greater than $5I_{\Delta n}$. The disconnection time should not exceed 40 ms (Regulation 643.8 refers).

Whilst product standard testing requires manufacturers' to subject RCDs to tests at half their rated residual operating current ($0.5I_{\Delta n}$), a half-rated test is not required by *BS 7671* to verify the effectiveness of an RCD, during initial verification or periodic testing.

Note: Testing RCDs at $0.5I_{\Delta n}$ might be useful when fault finding during maintenance or periodic inspection procedures.

RCD test procedure

Before an RCD test is conducted, the earth loop impedance of the circuit(s) protected by the RCD should be confirmed.

It should be noted that it is the RCD that is being subjected to a test and not the circuit or the load. Therefore, the instrument may be connected at any convenient point in the circuit(s) protected by the RCD, such as at a socket-outlet or at the outgoing terminals of the RCD. Any loads connected to the circuit(s) protected by the RCD should be disconnected for the purpose of the test (see Fig 15.12).

Generally, the RCD test instrument will have a facility to indicate whether or not polarity is correct. Testing should only proceed where correct polarity is being indicated.

The test is performed for both the positive and negative half-cycles of the supply waveform and the longest operating time obtained should be recorded in the appropriate box on the 'Schedule of circuit details and test results'. The disconnection time should be verified according to whether the RCD is being used to provide additional protection or fault protection.

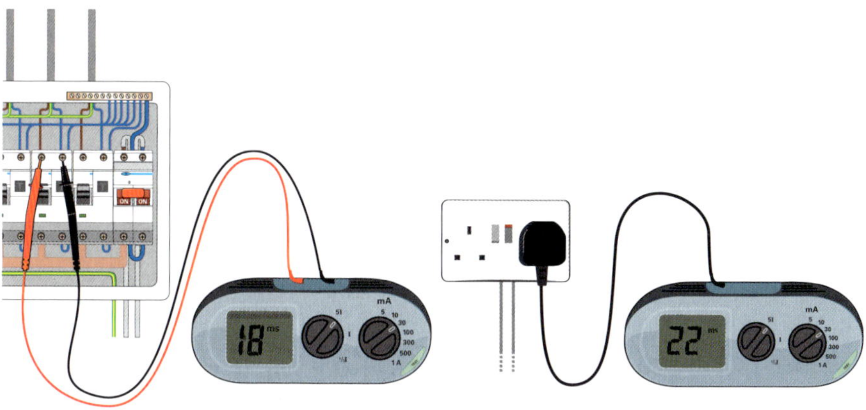

a) RCD test at a consumer unit

b) Test performed at a socket-outlet

Fig 15.12 RCD testing

Verification of the RCD test results

The results of any RCD testing should be verified according to the function of the RCD. Where an RCD is installed to provide additional protection, testing is required to verify that the RCD disconnects within 40 ms, as required by Regulation 415.1.1. Where an RCD is installed to provide fault protection, the testing should verify that the relevant disconnection times of Chapter 41 of *BS 7671* are met (Regulation 531.3.5.1 refers).

Table 15.3 below details the disconnection times of Chapter 41.

Table 15.3 Summary of operating times

Disconnection times of Chapter 41 of *BS 7671* for AC circuits (where $120\ V < U_0 \le 230\ V$)			
Circuit type / application	**Earthing system**	**Disconnection time (seconds)**	***BS 7671* reference**
Final circuits with a rated current not exceeding: (i) 63 A with one or more socket-outlets, and (ii) 32 A supplying only fixed connected current-using equipment	TN	0.4	411.3.2.2 Table 41.1
	TT	0.2	
Distribution circuits with rated current not exceeding 32 A	TN	5	411.3.2.3
	TT	1	411.3.2.4
Part of a reduced low voltage system	TN or TT	5	411.8.3

Following the electrical test procedure described above, each RCD should be operated by means of its integral test facility. This manual test is required to confirm that the device is responding to its design level of sensitivity and that all the mechanical parts are functioning.

For the benefit of the user of the installation, a notice should be fixed at or near each RCD in the installation stating that the integral test button should be operated at intervals not exceeding six months (Regulation 514.12.2 refers).

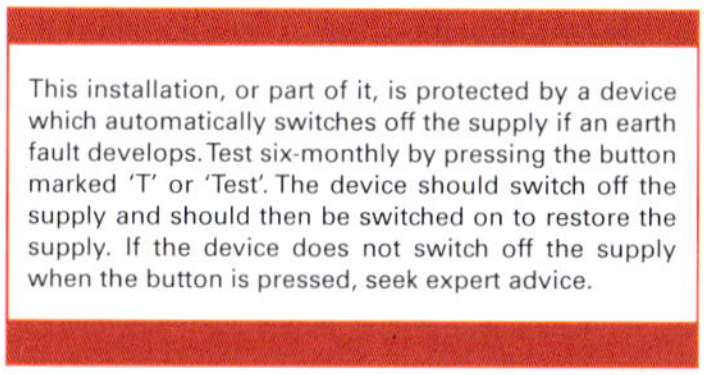

Fig 15.13 Six-monthly RCD test label

15.5.10 Functional testing

Regulation 643.10 requires it to be verified that the integral test button of an RCD functions correctly when pressed. This tests the ability of the mechanical components of the RCD to move freely and confirms that, for example, contacts have not "stuck".

Where an AFDD is installed the effectiveness of any manually operated test facility should also be verified.

It is also necessary to confirm that installed equipment such as switchgear, controlgear and interlock arrangements is properly and securely installed and has, where applicable, been correctly set up and adjusted ready for operation (Regulation 645.10 refers.)

Although coming under the heading of testing, it is likely that the above functional checks will, in practice, be achieved through inspection.

15.5.11 Certification and reporting

Regulation 644.1 requires an Electrical Installation Certificate to be issued on completion of the initial verification (that is, the inspection and testing) of the new electrical installation or of the changes to an existing installation.

Following completion of the periodic inspection of an existing installation Regulation 653.1 requires an Electrical installation Condition Report to be issued.

Both an Electrical Installation Certificate and an Electrical Installation Condition Report should be accompanied by sufficient Schedules of Inspections and Schedules of Test Results to cover the inspection and testing of the installation (see Regulations 644.3 and 653.2) and should be given to the person who ordered the work.

Where the electrical installation work is minor in nature - that is, it does not involve the provision of a new circuit, a Minor Electrical Installation Works Certificate (MEIWC) may be issued instead of an Electrical Installation Certificate (Regulation 644.4.201 refers).

However a MEIWC may only cover work carried out on a single circuit and so where minor works have been carried out on more than one circuit, a MEIWC must be issued for EACH circuit that has been worked on. Alternatively it might be better to issue an Electrical Installation Certificate in such cases.

Electrical Installation Certificates, Minor Electrical Installation Works Certificates and Electrical Installation Condition Reports should be produced and signed or otherwise authenticated by a person competent in such work (see Regulation 644.5 refers).

Detailed coverage on how to complete Electrical Installation Certificates, Electrical Installation Condition Reports and Minor Electrical Installation Works Certificates can be found in the following NICEIC publications:

- *Inspection, Testing and Certification*
- *Domestic Periodic Inspection, Testing and Reporting*

APPENDICES

A1 General

Final circuits, like all circuits of an electrical installation, have to comply with the requirements of *BS 7671*. The guidelines and examples of conventional circuit arrangements given in this Appendix are intended to satisfy the requirements of the following chapters of *BS 7671*:

- **Chapter 43** – Protection against overcurrent
- **Chapter 53** – Protection, isolation, switching, control and monitoring
- **Chapter 52** and **Appendix 4** – Current-carrying capacities of conductors and voltage drop.

The requirements of the other relevant chapters or sections of *BS 7671* also need to be considered and complied with, including:

- **Chapter 41** – Protection against electric shock
- **Section 434** – Protection against fault current
- **Chapter 54** – Earthing arrangements and protective conductors.

The guidance given in this Guide does not preclude other circuit arrangements where they meet the requirements of *BS 7671*.

A2 Floor area served

The floor area that can be served by a final circuit is dependent on the circuit loads either known or estimated, the diversity of those loads, the voltage drops that occur in the circuit, and the shock protection requirements of the circuit, such as the earth fault loop impedance (Z_S). The floor area served by the circuit is determined by the known or estimated load and should not exceed the value given in Table A1.

A single 30 A or 32 A ring final circuit may serve a floor area of up to 100 m^2.

Socket-outlets for washing machines, tumble dryers and dishwashers should be located so as to provide reasonable sharing of the load in each leg of the ring, or consideration should be given to separate circuits. A sufficient number of socket-outlets must be provided as required by Regulation 553.1.7.

Diversity between socket-outlets and permanently connected equipment has already been taken into account in Table A1 and no further diversity should be applied.

The cross-sectional area of the circuit protective conductor must be not less than that determined in accordance with Regulation Group 543.1.

Table A1 Final circuits using *BS 1363* socket-outlets and fused connection units

Type of circuit	Overcurrent protective device Rating (amperes)	Minimum conductor cross-sectional area (mm^2)		Maximum floor area served (m^2)
		Copper conductor thermoplastic (PVC) or thermosetting insulated cables (mm^2)	Copper conductor mineral insulated cables (mm^2)	
Ring	30 or 32	2.5	1.5	100
Radial	30 or 32	4	2.5	75
Radial	20	2.5	1.5	50

Note: Larger cable sizes could be required where the effective current-carrying capacity of the conductor has been reduced by other factors associated with the method of installation or external influences. Section 3 of Appendix 4 in *BS 7671* describes how such factors may be taken into consideration when determining an appropriate current-carrying capacity of a cable for continuous service (I_z), under particular conditions.

A3 Maximum demand

A final circuit should be designed so that its estimated maximum current demand does not exceed the rated current or current setting (I_n) of its overcurrent protective device, account being taken of the diversity of the loads envisaged. Where there is more than one final circuit serving multiple small loads, the permanently connected equipment and socket-outlets should be distributed amongst these circuits, where practicable, so as to maximize the load capacity available on each circuit for the socket-outlets.

Where a final circuit feeds *BS 1363* socket-outlets or fused connection units, the maximum current for which the circuit may be designed is restricted by the size, type and temperature of conductor that can be accommodated in the terminals of such accessories. *BS 1363* requires the incoming line and neutral terminals to be suitable for the connection, without special preparation, of one, two or three 2.5 mm^2 solid or stranded conductors, or one or two 4.0 mm^2 stranded conductors.

Furthermore, *BS 1363* accessories are designed for the connection of conductors operating at a temperature not exceeding 70 °C.

Thermosetting insulated, and some types of mineral insulated, cables can operate with conductor temperatures well above 70 °C. Where such cables are to be connected to socket-outlets and fused connection units complying with *BS 1363*, the cables are to be selected so that the conductor operating temperature is not in excess of 70 °C. In this respect, the conductor size of thermosetting insulated cable must not be less than the size that would have been required for thermoplastic (PVC) insulated cable.

Similarly, the conductor size of mineral insulated cable must not be less than that required for a mineral insulated cable intended to operate at not more than 70 °C (that is, a mineral insulated cable that is bare and exposed to touch or has an overall thermoplastic (PVC) covering). The restriction of conductor temperature to not more than 70 °C may also apply to other equipment in an installation (Regulation 512.1.5 refers).

A4 Limiting lengths for final circuits

Maximum permitted circuit lengths for some common situations are listed as follows:

Table A2 Radial final circuits supplying lighting

Table A3 Radial final circuits supplying accessories to *BS 1363*

Table A4 Radial final circuits supplying a single load

Table A5 Ring final circuits supplying accessories to *BS 1363*

In all cases, the tabulated data relates to the following conditions:

- The installation is supplied from a:
  - » TN-S system with a maximum external earth fault loop impedance (Z_e) of 0.8 Ω, or
  - » TN-C-S system with a maximum Z_e of 0.35 Ω, or
  - » TT system with a residual current device for fault protection. The circuit earth fault loop impedance in ohms is not to exceed the applicable value given in Table 41.5 of *BS 7671*, as required by Regulation 411.4.204.
- Maximum circuit cable lengths given for a TN-C-S system may also be applied to a TT system with RCD protection.
- The consumer unit and the main earthing terminal of the installation are located at the origin of the installation.
- The nominal supply voltage is 230 V single-phase AC
- The ambient temperature does not exceed 30 °C.

- 70 °C thermoplastic (PVC) insulated and sheathed flat type cable having copper conductors, conforming to Table 8 of *BS 6004* is used.

Note: The tables will also apply to cables conforming to *BS 7211* where they are operated at 70 °C rather than 90 °C.

- Circuit lengths are measured from the consumer unit to the most distant end of a radial circuit or the total route length for a ring final circuit.
- The maximum circuit cable length has been calculated to satisfy the requirements of *BS 7671* regarding voltage drop limitations and maximum disconnection times.
- In Table A2, a maximum voltage drop of 3 % of the 230 V nominal supply voltage has been allowed in the lighting circuits (Section 525 and Appendix 4 refer). See also Section 10.6 of this Guide.
- In Tables A3, A4 and A5, a maximum of 5 % of the 230 V nominal supply voltage has been allowed in the circuit (Section 525 and Appendix 4 refer). See also Chapter 9.5 of this Guide.
- Circuit cables are adequately spaced to give a current-carrying capacity (I_z) of at least the rated current (I_n) of the protective device (Regulation 433.1.1 refers). Where the protective device is a *BS 3036* fuse, I_z should be at least In divided by the 0.725 factor of Regulation 433.1.202.
- The grouping factors given in Table 4C1 of Appendix 4 in *BS 7671* allow for all the cables in the group to carry their design loads simultaneously and continuously. This is an unlikely event for circuits in household and similar premises, with the exception of cables to water heaters of storage tanks exceeding 15 litres capacity and electric space heating systems.
- With the exception of water heaters of storage tanks exceeding 15 litres and electric space heating, if the following rules are followed, derating for grouping is not necessary:

  i) Cables are not grouped when installed in or under insulation, that is, for installation methods 100, 101, 102, or 103.

  ii) Cables clipped direct (including in cement and plaster) are clipped side by side in one layer, preferably separated by one cable diameter.

  iii) Cables above ceilings are clipped to joists as per installation methods 100, 101, 102, and 103 of Table 4A2 of Appendix 4 in *BS 7671*.

Reduced cross-section protective conductors

In almost all household and similar circuits using 70 °C (PVC) insulated and sheathed flat cable with protective conductor conforming to Table 8 of *BS 6004*, the protective conductor cross-sectional area will be less than the live conductors. Here the requirement of Table 54.7 will not be met (for line conductors $S \leq 16\ mm^2$, protective conductors csa to be $S\ mm^2$) and so compliance with the adiabatic equation of Regulation 543.1.3 has been taken into account.

Where the protective device is a circuit-breaker to *BS EN 60898-1* or an RCBO to *BS EN 61009-1*, reference should also be made to Section A5 of this Appendix when determining protective conductor sizes.

Maximum disconnection time

A required disconnection time of 0.4 s has been presumed for all circuits having protective devices with a rated current (I_n) ≤ 32 A.

A required disconnection time of 5 s has been presumed for all circuits having protective devices with a rated current (I_n) > 32 A.

Where the rated current (I_n) of the protective device for a circuit ≤ 32 A, cable lengths for a disconnection time of 0.4 s correspond to the maximum values of Z_s given in Tables 41.2 and 41.3 of *BS 7671* (Regulations 411.4.201 and 411.4.202 refer).

Where the rated current (I_n) of the protective device for a circuit > 32 A, cable lengths for a disconnection time of 5 s correspond to the maximum values of Z_s given in Tables 41.2 and 41.3 of *BS 7671* (Regulations 411.4.202 and 411.4.208 refer).

Where no RCD is presumed (columns 6 and 8), circuit lengths are designed to limit earth fault loop impedance so that disconnection within 0.4 seconds is achieved for fuses (Table 41.2 of *BS 7671*) and instantaneous operation for circuit-breakers (Table 41.3 of *BS 7671*) in the event of a line-to-Earth fault.

Lighting circuits (Table A2)

The maximum route length of the radial circuit needs to be determined for the part of the circuit supplying the lighting load farthest from the consumer unit. The circuit route length is calculated by adding the appropriate circuit line and neutral conductor lengths (including a line conductor passing through a functional switching device), and then dividing the total by two. The maximum calculated value of circuit route length, between consumer unit and the farthest load is required to be equal to, or less than, the appropriate permitted maximum circuit route length (given in Table A2).

The circuit is assumed to have a load equal to the rated current (I_n) of the circuit protective device evenly distributed along the circuit. Where this is not the case, circuit lengths will need to be reduced where voltage drop is the limiting factor and halved where the load is all at the extremity.

The most onerous installation condition acceptable for the load and the device rating is presumed when calculating the limiting voltage drop. If the installation conditions are not the most onerous allowed (see column 4 of the tables), the voltage drop will not be as great as presumed in the table.

Radial final circuits supplying accessories to *BS 1363* (Table A3)

The circuit is assumed to have a load equal to the rated current (I_n) of the circuit protective device. 13 A of the load is assumed to be at the farthest end of the circuit from the consumer unit, with the remaining load being at 0.75 times L from the consumer unit, where L is the circuit length in metres between the consumer unit and the farthest end of the circuit.

Radial final circuits supplying a single load (Table A4)

Cables supplying whole house heating or water heaters of storage tanks exceeding 15 litres capacity should not be grouped unless appropriately derated.

Table A2: Maximum total route length for a 230 V radial final circuit supplying lighting loads in household premises using 70 °C thermoplastic (PVC) insulated and sheathed flat cable having copper conductors

	Protective Device	Cable size	Reference method	Maximum length (m)			
1	2	3	4	5	6	7	8
Rating (A)	Type	Line / cpc		$Z_e \leq 0.8\ \Omega$ TN-S		$Z_e \leq 0.35\ \Omega$ TN-C-S	
				RCD	No RCD	RCD	No RCD
5	*BS 1361* fuse *BS 3036* fuse *BS 88-3* fuse system C	1.0 / 1.0	103, 101, A 100, 102, C	71			
5	*BS 88-2.2* and *BS 88-6* fuse *BS EN 60898* circuit-breaker *BS EN 61009* RCBO - Type B *BS EN 60898* circuit-breaker *BS EN 61009* RCBO - Type C	1.5 / 1.0	103, 101, A 100, 102, C	108			
6	*BS 88-2.2* and *BS 88-6* fuse *BS EN 60898* circuit-breaker *BS EN 61009* RCBO - Type B *BS EN 60898* circuit-breaker *BS EN 61009* RCBO - Type C	1.0 / 1.0	103, 101, A 100, 102, C	59			
6	*BS 88-2.2* and BS 88-6 fuse *BS EN 60898* circuit-breaker *BS EN 61009* RCBO – Type B	1.5 / 1.0	103, 101, A 100, 102, C	90		90	
	BS EN 60898 circuit-breaker *BS EN 61009* RCBO – Type C			90	83	90	
10	*BS 88-2.2* and *BS 88-6* fuse *BS EN 60898* circuit-breaker *BS EN 61009* RCBO - Type B *BS EN 60898* circuit-breaker *BS EN 61009* RCBO - Type C	1.0 / 1.0	101, A, 100, 102, C	34		35	

Table A2 – continued

	Protective Device	Cable size	Reference method	Maximum length (m)			
1	2	3	4	5	6	7	8
Rating (A)	Type	Line / cpc		$Z_e \le 0.8\ \Omega$ TN-S		$Z_e \le 0.35\ \Omega$ TN-C-S	
				RCD	No RCD	RCD	No RCD
10	*BS 88-2.2* and *BS 88-6* fuse	1.5 / 1.0	103, 101, A 100, 102, C	52			
	BS EN 60898 circuit-breaker *BS EN 61009* RCBO - Type B						
	BS EN 60898 circuit-breaker *BS EN 61009* RCBO - Type C			51	41	52	52
16	*BS 88-2.2* and *BS 88-6* fuse	1.5 / 1.0	100, 102, C	33			
	BS 88-3 fuse system C						
	BS EN 60898 circuit-breaker *BS EN 61009 RCBO* - Type B						
	BS EN 60898 circuit-breaker *BS EN 61009 RCBO* - Type C			22	17	33	30
16	*BS 88-2.2* and *BS 88-6* fuse	2.5 / 1.5	101, A, 100, 102, C	53			
	BS 88-3 fuse system C						
	BS EN 60898 circuit-breaker *BS EN 61009* RCBO - Type B						
	BS EN 60898 circuit-breaker *BS EN 61009* RCBO - Type C			35	27	53	46

Table A3: Maximum total route length for a 230 V radial final circuit supplying accessories to *BS 1363* in household premises, using 70 °C thermoplastic (PVC) insulated and sheathed flat cable having copper conductors

	Protective Device	Cable size	Reference method	Maximum length (m)			
1	2	3	4	5	6	7	8
Rating (A)	Type	Line / cpc	(Note 2)	$Z_e \leq 0.8\ \Omega$ TN-S		$Z_e \leq 0.35\ \Omega$ TN-C-S	
				RCD	No RCD	RCD	No RCD
	BS 88-2.2 and *BS 88-6* fuse						
	BS 88-3 system C		A, 100, 102, C	35			
	BS 1361 fuse						
20	*BS 3036* fuse	2.5 / 1.5	NP	NP			
	BS EN 60898 circuit-breaker *BS EN 61009* RCBO - Type B		A, 100, 102, C	35			
	BS EN 60898 circuit-breaker *BS EN 61009* RCBO - Type C			19	14	35	34
	BS 88-2.2 and *BS 88-6* fuse		101, A, 100, 102, C		43		
	BS 88-3 system C			59	59	59	
	BS 1361 fuse		C		44		
20	*BS 3036* fuse	4.0 / 1.5		63	48	63	
	BS EN 60898 circuit-breaker *BS EN 61009* RCBO - Type B		101, A, 100, 102, C	59			
	BS EN 60898 circuit-breaker *BS EN 61009* RCBO - Type C			31	17	59	39
30	*BS 1361* fuse	4.0 / 1.5	C	43	NP	43	11
	BS 3036 fuse			NP			
	BS 88-2.2 and *BS 88-6* fuse				11		34
	BS 88-3 system C			42	7	42	30
32	*BS EN 60898* circuit-breaker *BS EN 61009* RCBO - Type B	4.0 / 1.5	C		31		42
	BS EN 60898 circuit-breaker *BS EN 61009* RCBO - Type C			NP		33	18

Notes:
1. NP = Not permitted 2. Any reference method less onerous than that in column 4 may be used

Table A4: Maximum total route length for a 230 V radial final circuit supplying loads such as a cooker, shower or heater at the end of the circuit using 70 °C thermoplastic (PVC) insulated and sheathed flat cable having copper conductors.

	Protective Device	Cable size	Reference method	Maximum length (m)			
1	2	3	4	5	6	7	8
Rating (A)	Type	Line / cpc	(Note 2)	$Z_e \leq 0.8\ \Omega$ TN-S		$Z_e \leq 0.35\ \Omega$ TN-C-S	
				RCD	No RCD	RCD	No RCD
15	*BS 1361* fuse	1.0 / 1.0	C	17			
	BS 3036 fuse		NP	NP			
15	*BS 1361* fuse	1.5 / 1.0	100, 102, C	26			
	BS 3036 fuse		NP	NP			
15	*BS 1361* fuse	2.5 / 1.5	101, A, 100, 102, C	43			
	BS 3036 fuse		100, 102, C	45			
15	*BS 1361* fuse	4.0 / 1.5	103, 101, A 100, 102, C	72			
	BS 3036 fuse		101, 100, 102, C	75			
16	*BS 88-2.2* and *BS 88-6* fuse	1.0 / 1.0	C	16			
	BS 88-3 system C						
	BS EN 60898 circuit-breaker *BS EN 61009* RCBO - Type B						
	BS EN 60898 circuit-breaker *BS EN 61009* RCBO - Type C			14		16	
16	*BS 88-2.2* and *BS 88-6* fuse	1.5 / 1.0	100, 102, C	24			
	BS 88-3 system C						
	BS EN 60898 circuit-breaker *BS EN 61009* RCBO - Type B						
	BS EN 60898 circuit-breaker *BS EN 61009* RCBO - Type C			22	17	24	

Table A4 - continued

	Protective Device	Cable size	Reference method	Maximum length (m)			
1	2	3	4	5	6	7	8
Rating (A)	Type	Line/cpc	(Note 2)	$Z_e \leq 0.8\ \Omega$ TN-S		$Z_e \leq 0.35\ \Omega$ TN-C-S	
				RCD	No RCD	RCD	No RCD
16	*BS 88-2.2* and *BS 88-6* fuse	2.5 / 1.5	101, A, 100, 102, C	40			
	BS 88-3 system C						
	BS EN 60898 circuit-breaker *BS EN 61009* RCBO - Type B						
	BS EN 60898 circuit-breaker *BS EN 61009* RCBO - Type C			35	27	40	
16	*BS 88-2.2* and *BS 88-6* fuse	4.0 / 1.5	103, 101, A, 100, 102, C	66			
	BS 88-3 system C						
	BS EN 60898 circuit-breaker *BS EN 61009* RCBO - Type B						
	BS EN 60898 circuit-breaker *BS EN 61009* RCBO - Type C			57	31	66	54
20	*BS 88-2.2* and *BS 88-6* fuse	2.5 / 1.5	101, A, 100, 102, C	31			
	BS 88-3 system C						
	BS 1361 fuse						
	BS 3036 fuse		NP	NP			
	BS EN 60898 circuit-breaker *BS EN 61009* RCBO - Type B		A, 100, 102, C	31			
	BS EN 60898 circuit-breaker *BS EN 61009* RCBO - Type C			19	14	31	

Notes:
1. NP = Not permitted
2. Any reference method less onerous than that in column 4 may be used

Table A4 - continued

	Protective Device	Cable size	Reference method	Maximum length (m)			
1	2	3	4	5	6	7	8
Rating (A)	Type	Line/cpc	(Note 2)	$Z_e \leq 0.8\ \Omega$ TN-S		$Z_e \leq 0.35\ \Omega$ TN-C-S	
				RCD	No RCD	RCD	No RCD
20	*BS 1361* fuse	4.0 / 1.5	101, A, 100, 102, C	53	44	53	
	BS 3036 fuse			57	48	57	
	BS 88-2.2 and *BS 88-6* fuse		C	53	48	53	
	BS 88-3 system C			53			
	BS EN 60898 circuit-breaker *BS EN 61009* RCBO - Type B		101, A, 100, 102, C	53			
	BS EN 60898 circuit-breaker *BS EN 61009* RCBO - Type C			31	17	53	39
20	*BS 88-2.2* and *BS 88-6* fuse	6.0 / 2.5	103, 101, A 100, 102, C	81	70	81	
	BS 88-3 system C			81	77	81	
	BS 1361 fuse		A, 100, 102, C	81	71	81	
	BS 3036 fuse			85	77	85	
	BS EN 60898 circuit-breaker *BS EN 61009* RCBO - Type B		103, 101, A 100, 102, C	81			
	BS EN 60898 circuit-breaker *BS EN 61009* RCBO - Type C			47	27	81	63
25	*BS 88-2.2* and *BS 88-6* fuse	2.5 / 1.5	C	26	23	26	
	BS EN 60898 circuit-breaker *BS EN 61009* RCBO - Type B			26			
	BS EN 60898 circuit-breaker *BS EN 61009* RCBO - Type C			6	5	26	24

Table A4 - continued

	Protective Device	Cable size	Reference method	Maximum length (m)			
1	2	3	4	5	6	7	8
Rating (A)	Type	Line/cpc	(Note 2)	$Z_e \leq 0.8\ \Omega$ TN-S		$Z_e \leq 0.35\ \Omega$ TN-C-S	
				RCD	No RCD	RCD	No RCD
25	*BS 88-2.2* and *BS 88-6* fuse	4.0 / 1.5	A, 100, 102, C	42	27	42	
	BS EN 60898 circuit-breaker *BS EN 61009* RCBO - Type B			42			
	BS EN 60898 circuit-breaker *BS EN 61009* RCBO - Type C			10	5	42	28
25	*BS 88-2.2* and *BS 88-6* fuse	6.0 / 2.5	101, A, 100, 102, C	64	43	64	
	BS EN 60898 circuit-breaker *BS EN 61009* RCBO - Type B			64			
	BS EN 60898 circuit-breaker *BS EN 61009* RCBO - Type C			16	9	64	45
30	*BS 1361* fuse	4.0 / 1.5	C	36	NP	36	11
	BS 3036 fuse		NP	NP			
30	*BS 1361* fuse	6.0 / 2.5	100, 102, C	53	NP	53	18
	BS 3036 fuse		NP	57	23	57	
30	*BS 1361* fuse	10.0 / 4.0	A, 100, 102, C	88	NP	88	29
	BS 3036 fuse		C	93	37	93	93
32	*BS 88-2.2* and *BS 88-6* fuse	4.0 / 1.5	C	33	11	33	30
	BS 88-3 system C				7		33
	BS EN 60898 circuit-breaker *BS EN 61009* RCBO - Type B				31	33	
	BS EN 60898 circuit-breaker *BS EN 61009* RCBO - Type C			NP		33	18

Note 2: Any reference method less onerous than that in column 4 may be used

Table A4 - continued

<table>
<tr><th></th><th>Protective Device</th><th>Cable size</th><th>Reference method</th><th colspan="4">Maximum length (m)</th></tr>
<tr><th>1</th><th>2</th><th>3</th><th>4</th><th>5</th><th>6</th><th>7</th><th>8</th></tr>
<tr><th rowspan="2">Rating (A)</th><th rowspan="2">Type</th><th rowspan="2">Line / cpc</th><th rowspan="2">(Note 2)</th><th colspan="2">$Z_e \leq 0.8\ \Omega$ TN-S</th><th colspan="2">$Z_e \leq 0.35\ \Omega$ TN-C-S</th></tr>
<tr><th>RCD</th><th>No RCD</th><th>RCD</th><th>No RCD</th></tr>
<tr><td rowspan="4">32</td><td>BS 88-2.2 and BS 88-6 fuse</td><td rowspan="4">6.0 / 2.5</td><td rowspan="4">A, 100, 102, C</td><td rowspan="3">49</td><td>19</td><td colspan="2">49</td></tr>
<tr><td>BS 88-3 system C</td><td>12</td><td>49</td><td>48</td></tr>
<tr><td>BS EN 60898 circuit-breaker
BS EN 61009 RCBO - Type B</td><td>49</td><td colspan="2">49</td></tr>
<tr><td>BS EN 60898 circuit-breaker
BS EN 61009 RCBO - Type C</td><td colspan="2">NP</td><td>49</td><td>29</td></tr>
<tr><td rowspan="4">32</td><td>BS 88-2.2 and BS 88-6 fuse</td><td rowspan="4">10.0 / 4.0</td><td rowspan="4">103, 101, A 100, 102, C</td><td rowspan="3">81</td><td>31</td><td colspan="2">81</td></tr>
<tr><td>BS 88-3 system C</td><td>20</td><td>81</td><td>78</td></tr>
<tr><td>BS EN 60898 circuit-breaker
BS EN 61009 RCBO - Type B</td><td>81</td><td colspan="2">81</td></tr>
<tr><td>BS EN 60898 circuit-breaker
BS EN 61009 RCBO - Type C</td><td colspan="2">NP</td><td>81</td><td>47</td></tr>
<tr><td rowspan="3">40</td><td>BS 88-2.2 and BS 88-6 fuse</td><td rowspan="3">6.0/2.5</td><td rowspan="3">C</td><td>40</td><td>NP</td><td>40</td><td>35</td></tr>
<tr><td>BS EN 60898 circuit-breaker
BS EN 61009 RCBO - Type B</td><td>40</td><td>27</td><td colspan="2">40</td></tr>
<tr><td>BS EN 60898 circuit-breaker
BS EN 61009 RCBO - Type C</td><td colspan="2">NP</td><td>29</td><td>17</td></tr>
<tr><td rowspan="3">40</td><td>BS 88-2.2 and BS 88-6 fuse</td><td rowspan="3">10.0 / 4.0</td><td rowspan="3">A, 100, 102, C</td><td rowspan="2">66</td><td>NP</td><td>66</td><td>56</td></tr>
<tr><td>BS EN 60898 circuit-breaker
BS EN 61009 RCBO - Type B</td><td>45</td><td colspan="2">66</td></tr>
<tr><td>BS EN 60898 circuit-breaker
BS EN 61009 RCBO - Type C</td><td colspan="2">NP</td><td>50</td><td>28</td></tr>
</table>

Notes:

1. NP = Not permitted
2. Any reference method less onerous than that in column 4 may be used

Table A4 - continued

<table>
<tr><th></th><th>Protective Device</th><th>Cable size</th><th>Reference method</th><th colspan="4">Maximum length (m)</th></tr>
<tr><th>1</th><th>2</th><th>3</th><th>4</th><th>5</th><th>6</th><th>7</th><th>8</th></tr>
<tr><th rowspan="2">Rating (A)</th><th rowspan="2">Type</th><th rowspan="2">Line/cpc</th><th rowspan="2">(Note 2)</th><th colspan="2">$Z_e \leq 0.8\ \Omega$ TN-S</th><th colspan="2">$Z_e \leq 0.35\ \Omega$ TN-C-S</th></tr>
<tr><th>RCD</th><th>No RCD</th><th>RCD</th><th>No RCD</th></tr>
<tr><td rowspan="3">40</td><td>BS 88-2.2 and BS 88-6 fuse</td><td rowspan="3">16.0 / 6.0</td><td rowspan="3">103, 101, A
100, 102, C</td><td rowspan="2">104</td><td>NP</td><td>104</td><td>86</td></tr>
<tr><td>BS EN 60898 circuit-breaker
BS EN 61009 RCBO - Type B</td><td>68</td><td colspan="2">104</td></tr>
<tr><td>BS EN 60898 circuit-breaker
BS EN 61009 RCBO - Type C</td><td colspan="2">NP</td><td>79</td><td>43</td></tr>
<tr><td rowspan="3">45</td><td>BS 1361 fuse</td><td rowspan="3">6.0 / 2.5</td><td>C</td><td>21</td><td rowspan="2">NP</td><td rowspan="2">35</td><td>NP</td></tr>
<tr><td>BS 88-3 system C</td><td>C</td><td>32</td><td>20</td></tr>
<tr><td>BS 3036 fuse</td><td>NP</td><td colspan="4">NP</td></tr>
<tr><td rowspan="3">45</td><td>BS 1361 fuse</td><td rowspan="3">10.0 / 4.0</td><td>A, 100,
102, C</td><td>36</td><td rowspan="3">NP</td><td rowspan="2">58</td><td>NP</td></tr>
<tr><td>BS 88-3 system C</td><td>A, 100,
102, C</td><td>54</td><td>33</td></tr>
<tr><td>BS 3036 fuse</td><td>C</td><td>62</td><td>62</td><td>31</td></tr>
<tr><td rowspan="3">45</td><td>BS 1361 fuse</td><td rowspan="3">16.0 / 6.0</td><td rowspan="2">A, 100,
102, C</td><td>57</td><td>NP</td><td>91</td><td>NP</td></tr>
<tr><td>BS 88-3 system C</td><td>86</td><td>NP</td><td>91</td><td>51</td></tr>
<tr><td>BS 3036 fuse</td><td>C</td><td>97</td><td>NP</td><td>97</td><td>47</td></tr>
</table>

Notes:
1. NP = Not permitted
2. Any reference method less onerous than that in column 4 may be used

Table A5: Maximum total route length for a 230 V ring final circuit in household premises using 70 °C thermoplastic (PVC) insulated and sheathed flat cable having copper conductors

	Protective Device	Cable size	Reference method	Maximum length (m)			
1	2	3	4	5	6	7	8
Rating (A)	Type	Line/cpc	(Note 2)	$Z_e \leq 0.8\ \Omega$ TN-S		$Z_e \leq 0.35\ \Omega$ TN-C-S	
				RCD	No RCD	RCD	No RCD
30	*BS 1361* fuse	2.5 / 1.5	A, 100, 102, C	111	NP	111	39
	BS 3036 fuse			111	49	111	
32	*BS 88-2.2* and *BS 88-6* fuse	2.5 / 1.5	A, 100, 102, C	106	41	106	
	BS 88-3 fuse system C			106	27	106	104
	BS EN 60898 circuit-breaker *BS EN 61009* RCBO - Type B			106			
	BS EN 60898 circuit-breaker *BS EN 61009* RCBO - Type C			NP		83	63
30	*BS 1361* fuse	4.0 / 1.5	101, A, 102	183	NP	183	45
	BS 3036 fuse		102, C	183	57	183	147
32	*BS 88-2.2* and *BS 88-6* fuse	4.0 / 1.5	101, A, 102	176	47	176	137
	BS 88-3 fuse system C			176	31	176	121
	BS EN 60898 circuit-breaker *BS EN 61009* RCBO - Type B		102, C	176	127	176	
	BS EN 60898 circuit-breaker *BS EN 61009* RCBO - Type C			NP		133	73

Notes:
1. NP = Not permitted
2. Any reference method less onerous than that in column 4 may be used

A5 Minimum permitted protective conductor sizes for energy limiting Class 3 circuit-breakers and RCBOs

The cross-sectional areas given in Table A6 overleaf have been derived using the following formula from Regulation 543.1.3 of *BS 7671*.

$$S = \frac{\sqrt{I^2t}}{k}$$

Where

S = cross-sectional area of protective conductor (mm^2)

I^2t = applicable energy let-through for type and class of protective device from Annex ZA of *BS EN 60898*

k = factor for type of protective conductor, taking into account

- the nature of the conductor
- conductor material, and
- insulation material.

A value of k of 115 has been used when producing Table A6. This *k* factor is taken from Table 54.3 of *BS 7671*. It is applicable for a cable having a copper protective conductor bunched with thermoplastic insulated copper conductors – as is the case for the 'twin and earth' type cables commonly used in domestic premises.

It is important that reference is made to the most appropriate part of the table in terms of the maximum prospective fault current that is likely to be present at the circuit-breaker or RCBO within the installation. In most cases, this will be significantly lower than the rated short-circuit capacity indicated on the circuit-breaker or RCBO.

It should be noted that the energy let-through values will be typically lower than those quoted in the product standards and, in some cases will give particular values for each rated current of protective device. It follows therefore that use of the manufacturer's data for the particular installed protective device will allow for the selection of a protective conductor of smaller csa than given in Table A6 in many cases.

Table A6 Minimum permitted protective conductor size for Energy limiting Class 3 circuit-breakers and RCBOs for fault current not exceeding 3 kA, 4.5 kA, 6 kA and 10 kA.

Rated current (A)	Circuit-breaker/RCBO type					
	Type B			Type C		
	I^2t max (A^2s)	Calculated minimum csa (mm^2)	Minimum available csa (mm^2)	I^2t max (A^2s)	Calculated minimum csa (mm^2)	Minimum available csa (mm^2)
3 kA fault current						
3 to 16	15000	1.064	1.5	18000	1.166	1.5
20 to 32	18000	1.166	1.5	22000	1.289	1.5
40	21600	1.277	1.5	26400	1.412	1.5
4.5 kA fault current						
3 to 16	25000	1.374	1.5	30000	1.506	2.5
20 to 32	32000	1.555	2.5	39000	1.717	2.5
40	38400	1.703	2.5	46800	1.881	2.5
6 kA fault current						
3 to 16	35000	1.626	2.5	42000	1.782	2.6
20 to 32	45000	1.844	2.5	55000	2.039	2.5
40	54000	2.020	2.5	66000	2.233	2.5
10 kA fault current						
3 to 16	70000	2.300	2.5	84000	2.52	4.0
20 to 32	90000	2.608	4.0	110000	2.884	4.0
40	108000	2.857	4.0	132000	3.159	4.0

Safe isolation

All persons carrying out electrical installation or maintenance work must be able to carry out the safe isolation of all or part of an electrical installation.

Regulation 12 of the *Electricity at Work Regulations 1989* deals with means of cutting off the supply and for isolation. The regulation requires that where necessary to prevent danger, suitable means (including, where appropriate, methods of identifying circuits) are made available:

- for cutting off the supply of electrical energy to any electrical equipment, and
- for the isolation of electrical equipment.

Isolation is not simply the operation of an isolating device. It is a procedure, the objective of which is to ensure that the supply is cut off and remains cut off for reasons of safety. The process of isolation can be broken down into a number of steps as necessary for the particular circumstances. As a minimum, the following safe working procedure should be adopted before commencing inspection (*HSG85: Electricity at Work safe working practices* gives detailed guidance on devising safe working practices for people who carry out work on or near electrical equipment).

The example illustrated overleaf shows the minimum steps required to isolate the final circuits supplied by a single-phase consumer unit. The consumer unit includes an isolator and circuit-breakers.

When circuits are protected by fuses enclosed in a consumer unit it may be necessary to isolate the whole consumer unit in some cases using the main switch.

Step 1

Check it is safe and acceptable (with the occupier/user) to isolate. If the isolator is an off-load device, remove the load.

Open the means of isolation for the circuit(s) to be isolated and secure the isolating device in the open position with a lock or other suitable means.

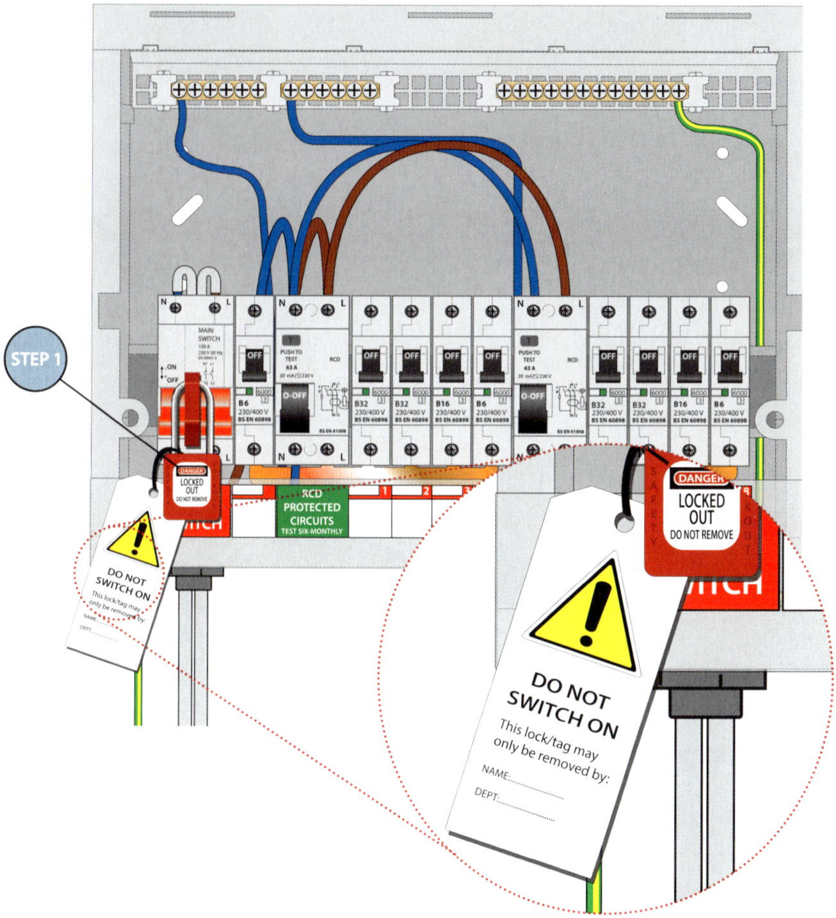

A label such as the one shown, should be attached to the padlock giving details of the person in charge of the key.

Step 2

Prove the correct operation of a suitable voltage detection instrument against a known voltage source, such as that illustrated. Guidance on voltage detection instruments is given in HSE Guidance Note *GS38 – Electrical test equipment for use on low voltage electrical systems.*

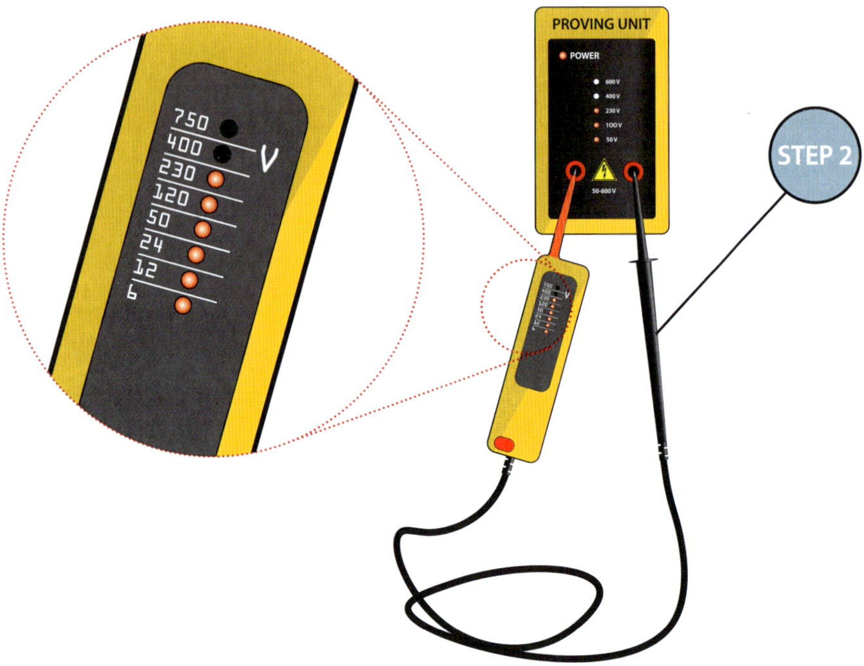

Step 3

Using a voltage detection instrument, check that there is no dangerous voltage present on any circuit conductor to be worked on or near to. It is important to confirm that conductors are not energised, for example, due to a wiring fault. Check terminal voltages between: (i) line and earth, (ii) line and neutral (as shown overleaf) and (iii) neutral and earth.

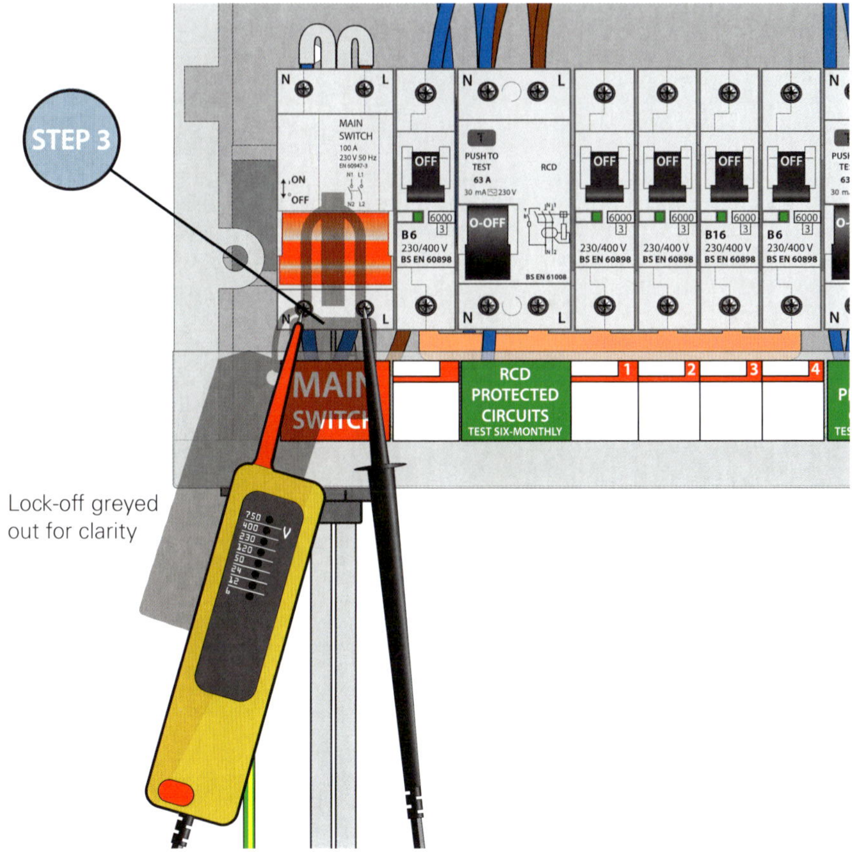

Step 4

Prove the voltage detection instrument again against the known source to check that it was functioning correctly when the circuit(s) was tested for the presence of voltage.

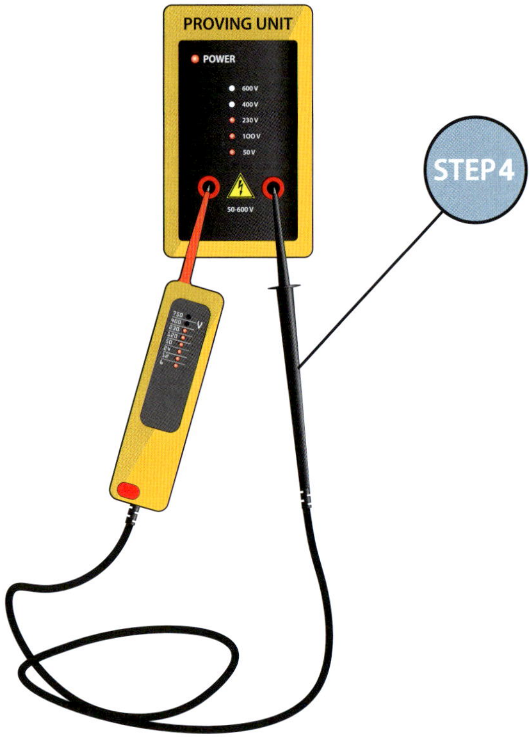

Notes:

1. In practice the equipment being worked on is likely to be remote from the consumer unit, for example, a socket-outlet located remotely from the means of isolation. In this case it is necessary to check that all the socket-outlet contact terminals are dead.
2. When checking for a voltage between an earth terminal and live (including neutral) terminals, the test probe should make contact with the earth terminal first, to reduce the risk of the remaining probe becoming live.

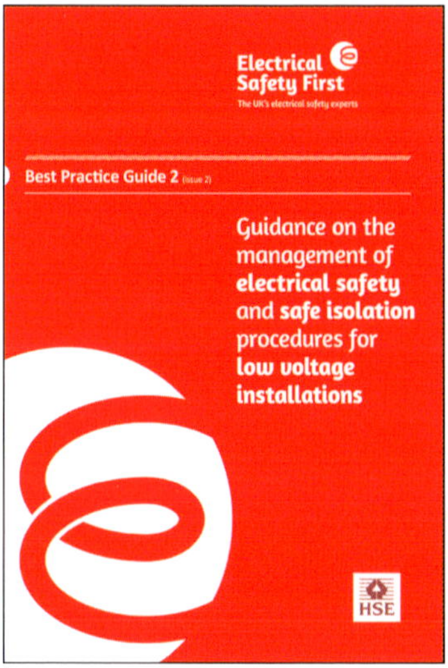

The Electrical Safety First Best Practice Guide 2, *Guidance on the management of electrical safety and safe isolation procedures for low voltage installations* has been produced by Electrical Safety First in conjunction with the Health and Safety Executive (HSE) and other organisations. Its purpose is to provide practical guidance for employers and the self-employed on safe isolation procedures to be followed when working on low voltage electrical installations.

The Guide can be downloaded from Electrical Safety First's website **www.electricalsafetyfirst.org.uk**

Guidance on the 'isolation procedure' is also provided in the NICEIC and ELECSA *Pocket Guide 5 – Guide to isolation procedure.*

All NICEIC and ELECSA Pocket Guides are available to download from the NICEIC and ELECSA website: **www.niceic.com and www.elecsa.co.uk**.

Resistance of copper conductors

Information on the resistance of copper and aluminium conductors is given in *BS EN 60228: 2005 Conductors of insulated cables.*

For ease of reference, Table A1 gives resistance per metre values of copper conductors of single and multicore cables, whether of solid or stranded construction, at 20 °C, in units of milliohms per metre (mΩ per m).

The table is based on data contained in Tables 1 and 2 of *BS EN 60228*

Table A1 – Maximum resistance of conductors at 20 °C

Conductor cross-sectional area (mm^2)	Resistance per metre of copper conductor (mΩ per m)
1.0	18.1
1.5	12.1
2.5	7.41
4.0	4.61
6.0	3.08
10	1.83
16	1.15
25	0.727
35	0.524
50	0.387

Data courtesy of the British Standards Institution

Conductor resistance values at a conductor temperature other than 20 °C

Resistance values shown in Table A1 may be corrected to other resistance values, for conductor temperatures other than 20 °C, by using the formula:

$$R_x = R_{20}(1 + (0.004 \times (T_x - T_{20})))$$

Where:

R_x is the corrected conductor resistance value, in mΩ, at x °C

R_{20} is the conductor resistance value at 20 °C, in mΩ, (see Table A1 for values)

0.004 is the approximate resistance temperature coefficient for copper or aluminium

T_x is the x °C conductor temperature

T_{20} is the 20 °C conductor temperature.

Calculation of conductor resistance at 30 °C

To calculate the resistance of a 6 mm^2 cross-sectional area (csa) copper conductor at a temperature of 30 °C:

T_{30} = 30 °C

R_{20} = 3.08 mΩ for a 6 mm^2 csa copper conductor, at 20 °C.
(The value of 3.08 mΩ is taken from column 2 of Table A1.)

Substituting the above values in the previous formula gives:

$R_{30} = R_{20}(1 + (0.004 \times (T_{30} - T_{20})))$

$R_{30} = 3.08 (1 + (0.004 \times (30 - 20))) = 3.08 (1 + (0.004 \times 10))$

$R_{30} = 3.08 \times (1 + 0.04)$

$R_{30} = 3.08 \times 1.04$

$R_{30} = 3.2$ mΩ

It can be seen from the above calculation that the original value of conductor resistance at 20 °C is increased by a multiplication factor of 1.04 for a conductor temperature of 30 °C. (30 °C is the 'assumed initial conductor temperature' used in Table 54.2 of *BS 7671*.)

Conductor resistance at 70 °C

By carrying out a calculation similar to that shown in the previous exercise, it can be shown that the resistance of a conductor at 70 °C is 1.2 times its resistance value at 20 °C. (70 °C is the 'assumed initial conductor temperature' used in Table 54.3 of *BS 7671* for 70 °C thermoplastic insulated cables.)

These commonly used conductor temperature correction factors are given in Table A2, for ease of reference (since they are used, and referenced, elsewhere in this book).

Further information on conductor temperature correction multiplication factors

Temperature correction factors for conductor resistance to correct measured resistance t °C are given in Table A.1 of *BS EN 60228* and exact formulae for temperature correction factors are stated in Annex B of *BS EN 60228*.

Table A2 – Resistance temperature multiplication correction factors for protective conductors with thermoplastic insulation, which are protected by standard overcurrent devices (given in Tables 41.2, 41.3 and 41.4 of *BS 7671*)

Protective conductor with thermoplastic (pvc) insulation	Initial conductor temperature (see Tables 54.2 and 54.3 of *BS 7671*)	20 °C conductor resistance correction multiplication factor (for applying to values in Table A1)
Applies where conductor is not incorporated in a cable or bunched with cables, or are conductors in contact with cable covers	30 °C	1.04
Applies where conductor is incorporated in a cable or bunched with other cables	70 °C	1.2